WATER RESOURCES STRATEGIES TO INCREASE FOOD PRODUCTION IN THE SEMI-ARID TROPICS

WITH PARTICULAR EMPHASIS ON THE POTENTIAL OF ALLUVIAL GROUNDWATER

WATER RESOURCES STRATEGIES TO INCREASE FOOD PRODUCTION IN THE SEMI-ARID TROPICS

WITH PARTICULAR EMPHASIS ON THE POTENTIAL OF ALLUVIAL GROUNDWATER

DISSERTATION

Submitted in fulfillment of the requirements of
the Board for Doctorates of Delft University of Technology
and of the Academic Board of the UNESCO-IHE
Institute for Water Education
for the Degree of DOCTOR
to be defended in public on
Thursday, 12 September 2013, at 12.30 hrs
in Delft, the Netherlands

by

David LOVE

Bachelor of Science, University of Zimbabwe
Bachelor of Science Honours and Masters of Science, University of
Stellenbosch, South Africa

born in Lusaka, Zambia.

This dissertation has been approved by the supervisors:
Prof. dr. S. Uhlenbrook
Prof. dr. ir. P. van der Zaag

Composition of Doctoral Committee:

Chairman	Rector Magnificus TU Delft
Vice-Chairman	Rector UNESCO-IHE
Prof. dr. S. Uhlenbrook	UNESCO-IHE / Delft University of Technology
Prof. dr. ir. P. van der Zaag	UNESCO-IHE / Delft University of Technology
Prof. dr. ir. H.H.G Savenije	Delft University of Technology
Prof.dr.ir. N.C van de Giesen	Delft University of Technology
Prof dr. C. de Fraiture	UNESCO-IHE / Wageningen University
Prof. dr. D. Mazvimavi	University of the Western Cape, South Africa
Dr. ir.T.N. Olsthoorn	Delft University of Technology, reserve member

CRC Press/Balkema is an imprint of the Taylor & Francis Group, an informa business

Published by:
CRC Press/Balkema
PO Box 11320, 2301 EH Leiden, The Netherlands
e-mail: Pub.NL@taylorandfrancis.com
www.crcpress.com - www.taylorandfrancis.com

ISBN 978-1-138-00142-8

Acknowledgements

In late 2003, I was working as a lecturer in the Geology Department at the University of Zimbabwe, when WaterNet (of which the Department is a member) invited us to participate in preparing a proposal to the Challenge Program on Water and Food for a transdisciplinary water research project in the Limpopo Basin. The proposal was successful, and I was subsequently recruited as a PhD fellow and as a part-time project coordinator at WaterNet.

WaterNet is without doubt one of the most dynamic organisations that I have worked for. It is also modern, delivering education, research and outreach through teams from many different institutions (Mode 2 Knowledge Production *per* Gibbons) that make up the network. This is especially important for our sub-region, as many countries in SADC have small populations and thus do not always have world class human resources in some areas of expertise – but it can always be found by sharing with neighbours.

Our PN17 project *Integrated Water Resources Management for Improved Rural Livelihoods*, set out to demonstrate that improving water management at any scale improves people's livelihoods. I hope that my work has in its own way contributed to this and provides the Mzingwane Catchment Council with some useful insights and tools for the management of their sub-basin. I was lucky to participate in such an interesting project and work with such great colleagues.

The science has been fascinating, dealing with hydrogeology, hydrology, crop science and development – and interacting with and learning from colleagues in all these fields. I was priviledged to receive awards for two of my papers: the Tison Award from the International Association of Hydrological Sciences and the Phaup Award from the Geological Society of Zimbabwe.

This has been a personal journey for me, not just the science and the wonderful communities we worked with. The initiation of the project coincided with my marriage to Faith, and discovering that her eldest brother, Phanuel Ncube, chaired the Mzingwane Catchment Council - the water management authority that WaterNet planned to work with. So this journey has been a fundamental part of my family life as well as my professional life.

To my promoters and supervisors, my sincere appreciation for your guidance, mentorship, support and endless patience. Stefan Uhlenbrook and Pieter van der Zaag provided the overall scientific guidance and the confidence and support to keep going through the long years of this research, especially when my morale was low. At no time were you too busy to help, correct and guide. Richard Owen provided the detailed guidance on hydrogeology and encouraged the growth of my love for the alluvial aquifers of the hotter and drier parts of our sub-continent – as well as giving me support in matters personal and spiritual. Steve Twomlow guided me through the science of semi-arid lands, field discipline and support at ICRISAT Matopos Research Station.

This work is part of a trans-disciplinary trans-institutional project, and I wish to thank my many colleagues in the Limpopo Basin, including my PhD co-fellows, *sekuru* Walter Mupangwa, Collin Mabiza, Paiva Munguambe and Manuel

Magombeyi. At the Zimbabwe National Water Authority: Tommy Rosen, Elisha Madamombe, Charles Sakuhuni, and my brother-in-law and longstanding chairman of the Mzingwane Catchment Council, *tezvara* Phanuel Ncube. At ICRISAT, Andre van Rooyen, John Dimes and my *mai gurus* Bongani Ncube and Sifiso Ncube, and the drivers with whom I spent so much time in the field, especially Mr Masuku, Mr Manyani, Mr Mlotshwa and Mr Mpofu. At the University of Zimbabwe, Hodson Makurira, Innocent Nhapi and Aiden Senzanje, and the technicians who also spent time with me in the field, especially Farai Zihanzu, Percy Sena and Douglas Maguze. At WaterNet, Lewis Jonker, Johan Rockström, Themba Gumbo, Bongani Ncube (again), *mkoma* Washy Nyabeze, Nick Tandi, Martha Hondo, Moriah Makopa, Admire Mutowembwa, Jean-Marie Kieshye-Onema and Rennie Munyayi. At the Dabane Trust, Steve Hussey and Ekron Nyoni.

In the Netherlands, I benefitted greatly from interactions with many staff and PhD fellows, especially Gerald Corzo Perez, with the time developing HBVx together and with whom I shared the Tison Award - and also Marloes Mul, Hodson Makurira (again), Collin Mabiza (again), Marieke de Groen, Ilyas Masih, Jeltsje Kemerink and Ann van Griensven.

In South Africa, to my new colleagues at Golder: Keretia Lupankwa, Koovila Naicker, Nico Bezuidenhout and Gerhard van der Linde, for your support in the last months of this long journey, and Washy Nyabeze (again), Marieke de Groen (again), Themba Gumbo (again) and Bongani Ncube (again) for your advice and encouragement.

To the many Masters students from University of Zimbabwe, UNESCO-IHE, TU Delft and University of Twente, it was a pleasure supervising you and your contributions to this work, direct and indirect is highly valued.

To my field assistants, Tius Sibanda, Sanelisiwe Sibanda, Tutanang Nyati, Patrick Nyati, Regis Mukwane, Brighton Sibanda, Daniel and Saddam Mkwananzi and my *mai ninis* Sipatisiwe and Sipetokuhle Ncube your contribution was also vital. Additional rainfall data was obtained with the help of the Beitbridge Bulawayo Railway Company, Mazunga Safaris, Tod's Guest House and the communities of Zhulube (Insiza), Fumukwe and Manama (Gwanda), Maranda, Nemangwe and Chengwe (Mwenezi) and Dendele, Tongwe, Malala and Masera (Beitbridge).

The assistance of *baba mkuru* Hayi Mpofu (Zhulube Irrigation Scheme), Felix Whinya (Zhovhe Dam), Tod's Guest House, Paul Bristow and Rob Smith (Mazunga), and the District Administrators and Rural District Councils of Insiza, Gwanda, Mwenezi and Beitbridge Districts has been essential and is gratefully acknowledged.

This thesis is an output of the CGIAR Challenge Program on Water and Food Project 17 "Integrated Water Resource Management for Improved Rural Livelihoods: Managing risk, mitigating drought and improving water productivity in the water scarce Limpopo Basin", led by WaterNet, with additional funding provided by the International Foundation for Science (Grant W4029-1). Work in the field was also supported by ICRISAT Matopos, the Dabane Trust, the University of Zimbabwe Department of Geology and World Vision Insiza ADP. Piezometers were designed and manufactured by the Dabane Trust and installed with their assistance. The opinions and results presented are those of the author and do not necessarily represent the donors or participating institutions.

Reference discharge data were provided by the Zimbabwe National Water Authority and rainfall data by the Department of Meteorological Services, Zimbabwe Ministry of Environment and Tourism. Additional data were also availed by the Zhovhe Water Users Association. Jan Seibert kindly provided the HBV light 2 code.

To my dear wife Faith, for your endless patience, love and support, and to our two children born during this work, Kathleen Taboka and James Robert Langanani, a big thank you for bearing with me when I was away, accompanying me some of the time and supporting me all the way through.

Finally, I give honor and glory to God, for all His gifts to me, and especially for the gifts of intellect and science, that we may know more of the workings of this World.

David Love, Midrand, August 2013

Summary

A number of hydroclimatic and institutional factors converge to emphasise the need for investment in water management and water resources modelling in southern Africa. Water demand continues to rise, as urban areas expand and as agricultural water demand rises to meet the millennium development food security goals. In average years, water demand (principally from agriculture and urban areas) is in a precarious balance with available water resources, with major deficits and severe food insecurity being recorded during droughts. Access to water is limited by actual scarcity, availability and affordability of water storage and appropriate abstraction technology and water allocation practices. This study shows that water resource availability in the northern Limpopo River Basin (i.e. the portion of the Limpopo Basin located in Zimbabwe, also known as the Mzingwane Catchment) has declined over the last 30 years, both in terms of total annual water available for storage (i.e. declines in annual rainfall, annual runoff) and in terms of the frequency of water availability (i.e. declines in number of rainy days, increases in dry spells, increases in days without flow). Furthermore, a number of climate change models predict that southern Africa shall experience significantly reduced precipitation and runoff over the next fifty years. Simulation modeling suggests a more than proportional decline in runoff and water for productive use.

Changes in water and land use strategies can have significant effects on water resources. Increases in irrigation, and construction of reservoirs, have obvious effects on river systems, but rainfed agriculture and land use changes, whilst not necessarily exhibiting demand for surface or groundwater, can exert a strong influence on runoff generation. In this context, there is a clear requirement for water resources modelling to support integrated water resources management planning in order to balance food security, other economic needs and the needs of the environment in the allocation and development of blue water flows.

This study seeks to model water resources at river basin scale in order to quantify the effect of different water and land use strategies and hydrological and climatic conditions on water resources availability.

An extended version of the HBV light rainfall-runoff model was developed (designated HBVx), introducing an interception storage and with all routines run in semi-distributed mode via visual basic macros in a spreadsheet. This was used to characterise the response of meso-catchments in the study area to rainfall, in terms of the production of runoff vs. the interception, transpiration and evaporation of water. This is important in small semi-arid catchments, where a few intense rainfall events may generate much of the season's runoff. HBVx was regionalised across 19 meso-catchments and satisfactorily models the ephemeral surface flow and the minimal baseflow from deep groundwater in semi-arid meso-catchments. Meso-catchments in the study area are characterised by high levels of interception, slow infiltration and percolation and moderate to fast overland flow.

The alluvial aquifers that form the beds of sand rivers are perennial in ephemeral rivers, largely protected from evaporation and normally of good quality. The northern Limpopo Basin has erratic and unreliable rainfall and very low mean annual runoff. Alluvial aquifers thus present an attractive option for water management: firstly for conjunctive use with surface water for the storage of water in ephemeral sand rivers and secondly as a sustainable alternative to surface water use. The water supply potential of a case study alluvial aquifer was evaluated using field observations and the finite difference groundwater flow model, MODFLOW. The behaviour of the aquifer under higher seepage, and climate change and development scenarios was also modelled. This showed that alluvial aquifers of this scale are suitable for use for domestic and livestock water supply and the irrigation of small gardens.

A remote sensing approach was used to identify and map 1,835 km of alluvial aquifers in the northern Limpopo Basin.

Using a combination of field and laboratory investigations, remote sensing and existing data, the Lower Mzingwane valley was modelled successfully using the spreadsheet-based model WAFLEX, with a new module incorporated to compute the water balance of alluvial aquifer blocks. The model performed adequately and produced a trend in findings on alluvial aquifer behaviour comparable to published field studies. It thus provides a flexible tool for the evaluation of alluvial aquifers on large, lowland rivers and can provide useful information for planning purposes from limited data.

HOWSIT, a coupled, spreadsheet-based model is used to evaluate land/water scenarios and livelihood intervention strategies in the Insiza Catchment, Limpopo Basin. The model integrates HBVx and WAFLEX in a spreadsheet and evaluates climate change and intervention responses such as changing crop choice and irrigation method, conservation agriculture, new irrigated lands and crop-livestock integration through their effects on soil/water dynamics, runoff generation, aquifer recharge and water allocation. The coupled HOWSIT model performs moderately well in this context, but shows significant sensitivity to soil parameters, higher than the change simulated in many development scenarios. Uncertainty in water resources and allocation planning can be decreased through the use of conservative assumptions, with which HOWSIT can give robust minimum values for water allocation purposes.

The greatest benefit for the least impact comes from the strategies providing for better land and soil/water management. Changing from maize to small grains and to production of livestock fodder show clearly excellent results. Conservation agriculture has potentially a similar impact. Small dams are a key resource to rural communities, and change to Multiple Use Systems through abstraction of water for irrigation has clear benefits locally, although the downstream impact is relatively high for this benefit, unless drip irrigation is practiced. Better management of existing large dams, together with conjunctive reservoir use (where water is released from an upstream reservoir, when the water level in a downstream reservoir falls below a critical level) could increase the productive use of water and thus significantly improve livelihoods without the need for the construction of new reservoirs.

There is great potential for the exploitation of alluvial aquifers for irrigation water supply. A total of 6,740 ha of land can be irrigated by exploitation of alluvial aquifers, mainly in the lowland rivers and tributaries. This irrigation would be decentralised, owned and operated at household level and the benefits would have the potential to reach a much larger proportion of the population than is currently served.

This study concludes with a series of recommendations on water resources development and agricultural water management. Critically, expansion of irrigation in the Mzingwane Catchment should focus on the conjunctive use of surface water and groundwater and thorough investigation of possible water supply from alluvial aquifers should be part of the scoping exercise when considering the possible construction of these new large reservoirs. Functional and operational decentralisation of smallholder irrigation, using strips of land adjacent to rivers rather than concentrating only on large schemes, should allow for low cost abstraction systems and household scale control of water use, potentially leading to better household investment decisions and improved access to water for women

Priority should be given in extension and outreach to changes in rainfed cropping which can benefit a greater proportion of the population without significant downstream impact on water resources. This should include changing crop from maize to more drought tolerant crops, specifically sorghum and millet, and the production of cattle fodder.

These approaches of conjunctive water use and changes in rainfed cropping need to be built into the extension curriculum and the training of catchment councillors. Increased knowledge, improved technology transfer and better allocation models and practices can lead to more equitable and productive use of water in the semi-arid lands.

Table of Contents

1. Introduction

1.1. Background to this study

1.1.1. Challenges for Water Resources in Southern Africa

Southern Africa faces the twin challenges of declining water resource availability and rising water demand.

Rainfall in south-eastern Africa is temporally and spatially intermittent (Unganai and Mason, 2002). Annual rainfall for a single site can vary by up to 1000 mm a^{-1} from year to year – although a drought year may record less than 250 mm a^{-1}, such as the 2004–2005 season in the Limpopo Basin (Love et al., 2006a). Rainfall variability is strongly influenced by the coupled ocean-atmosphere El Niño – Southern Oscillation phenomenon (ENSO) (Trenberth *et al.*, 2007). Positive ENSO anomalies generally result in reduced rainfall in the region and are becoming more common (Makarau and Jury, 1998; Alemaw and Chaoka, 2006). Furthermore, there has been a general decline in rainfall in southern Africa since 1961 (New *et al.*, 2006), with the period 1986-1995 being the driest decade of the twentieth century (Trenberth *et al.*, 2007).

General circulation models developed with the Intergovernmental Panel on Climate Change Special Report on Emission Scenarios (IPCC SRES) scenarios suggest annual rainfall in south-eastern Africa will decline further under the impact of global warming (Desanker and Magadza, 2001; Christensen et al., 2007; Andersson *et al.*, 2011), especially in Botswana and Zimbabwe (Engelbrecht *et al.*, 2011). This is expected to be between 10% and 20% below the 1900-1970 averages by 2050 (Milly et al., 2008) or by up to 10% below the 1980-1999 averages by 2099 (Christensen et al., 2007).

Declines in rainfall as discussed above may translate to more than proportional declines in discharge due to non-linear processes, including for example interception thresholds. By the 2050s, in the south-eastern Africa, runoff is expected to decline by between 10% and 40% compared to 1961-1990 averages (De Wit and Stankiewicz, 2006). Thus water resource availability is declining as rainfall and runoff decrease (Love *et al.* 2010b) and is likely to do so further as the impact of climate change on water resource availability is felt. Other changes such as a delay in the onset and early cessation of the rainy season, and an increase in the severity of droughts can also be expected (Shongwe *et al.*, 2009).

Water demand is increasing, mainly from irrigated agriculture, which must expand to meet food needs, but also due to the water supply requirements of rapidly-growing urban areas (Ncube *et al.*, 2010; Van der Zaag and Gupta, 2008). Demand for irrigation water is likely to rise as climate change reduces dryland crop production (Stige *et al.*, 2006) – but also as agriculture is a major priority for economic growth in Sub-Saharan Africa (Commission for Africa, 2005). Expansion of irrigated agriculture and development of groundwater and alternative water sources are priorities for southern Africa in the SADC climate change strategy (SADC, 2011).

1.1.2. *Water Resources Management and Livelihood Strategies*

The changing regional and global trends in climate and discharge discussed above, and their influence on water resource availability, will increase livelihood risk. It has been shown that household food security in southern Africa is highly vulnerable to climate stress (Archer *et al.*, 2007). Already in much of southern Africa there is a precarious balance between available water resources and water demand as a result of generally low conversion of rainfall to runoff and potential evaporation exceeding rainfall (e.g. Farquharson and Bullock, 1992; Mazvimavi, 2003). Frequently, the water yield from the developed surface water resource falls short of the demand, deficits being more evident during the frequent droughts (e.g. Nyabeze, 2004). Furthermore, some catchments, especially within the Limpopo Basin, are already over-committed (Kabel, 1984; Basson and Rossouw, 2003), leading to water stress: a high ratio of water withdrawal or water use to discharge (Vörösmarty *et al.*, 2000). Changes of this nature constitute a major challenge to water resources management (Milly et al., 2008).

The first Millennium Development Goal (MDG) aims to eradicate extreme poverty and hunger. Target 2 of Goal 1 is to halve, between 1990 and 2015, the proportion of people who suffer from hunger (UN Millennium Project, 2005a). This is extremely important in southern Africa, where food security has become increasingly problematic in the last half-century. In this context, governments, development agents, NGOs and individual farmers are all developing strategies and interventions to improve food security and rural livelihoods. Most of these strategies and interventions have an effect on the water cycle, through increasing the demand for blue water, through changing green water use or both (Falkenmark and Rockström, 2004; Love *et al.*, 2006a; Moyo *et al.*, 2006; Mupangwa *et al.*, 2006; Hanjra and Gichuki, 2008; Van der Zaag, 2009; Vidal *et al.*, 2010). The challenge and its response thus exist at a nexus of stressed water resource availability. Nowhere is this clearer than in the semi-arid lands, such as the Limpopo Basin.

It is within this context that the Consultative Group on International Agricultural Research (CGIAR) launched the Challenge Program on Water and Food (CPWF), in 2002 to increase the resilience of social and ecological systems through better water management for food production. Project 17, led by WaterNet, sought to contribute to improved rural livelihoods of poor smallholder farmers through the development of an Integrated Water Resource Management (IWRM) framework for increased productive use of water flows (Ncube *et al.*, 2010).

During the scoping of this study, a farmer from a community in Gwanda South made the comment that "*we see the water flowing away down the big rivers every year, and yet we receive none of this and our crops fail.*" Whilst there is a role for large-scale industrial projects such as large dams and huge irrigation schemes, to increase water resource availability, there is considerable scope at the smaller scale. This thesis examines the water resources aspects of approaches that households and small communities can themselves take to improve their access to water, thus increasing household water resource availability, and potentially, food production.

This thesis contributes to CPWF Project 17 through investigating the hydrology, hydroclimatology and hydrogeology of the study catchments and then using modelling to evaluate the benefits and impacts of various water resources management strategies.

1.1.3.   The Role of Water Resources Modelling

The first step in understanding water resource availability in a catchment is characterising the response of a catchment to rainfall, in terms of the production of runoff vs. the interception, transpiration and evaporation of water. This is particularly important in semi-arid catchments, where a few intense rainfall events may generate much, or sometimes most, of the season's runoff (e.g. Lange and Leibundgut, 2003) and where spatial and temporal variability of rainfall can be high (e.g. Unganai and Mason, 2002). An understanding of the hydrological processes involved in a catchment is a basic requirement for integrated water resources management planning (e.g. Uhlenbrook et al., 2004) as well as for understanding the hydrological impact of changes in land use and agricultural water management. Such changes can affect all components of the water balance in a catchment: changes in land cover affect evaporation, interception and transpiration; soil processes are affected by water consumption by plants and physical changes to the soil structure from roots – and these processes in turn affect runoff generation and percolation to groundwater (Uhlenbrook, 2007).

In southern Africa, where environmental and water stress is increasing (Nyabeze, 2004; Sivakumar et al., 2005), this type of understanding is essential in building resilience to large or catastrophic environmental changes and in developing trade-offs between food and economic production and ecosystem services (Falkenmark et al., 2007). It is also important for addressing broader humanitarian and development needs, through the many water intensive interventions that have been proposed by development agencies and projects (Love et al., 2006a). Given the high variability in rainfall and soil types in Sub-Saharan Africa, van der Zaag (2009) argues for location-specific interventions, and these will require a proper understanding of the local catchments.

Another important aspect of water resource availability is groundwater – surface water interactions. The sandy beds of rivers host alluvial aquifers. An alluvial aquifer can be described as a groundwater unit, generally unconfined, that is hosted in laterally discontinuous layers of sand, silt and clay, deposited by a river in a river channel, banks or flood plain (Barker and Molle, 2004). This study focuses on the alluvial aquifers in river channels. Because of their shallow depth and their vicinity to the streambed, alluvial aquifers have an intimate relationship with surface flow, which is the main source of aquifer recharge. Indeed it can be argued that groundwater flow in alluvial aquifers is an extension of surface flow (Mansell and Hussey, 2005; Love et al., 2010c). In arid and some semi-arid areas, alluvial aquifer recharge may occur only after high discharge peaks from heavy rainfall events (Lange and Leibundgut, 2003; Lange, 2005; Matter et al., 2005; Benito et al., 2010) and full recharge normally occurs early in the rainy season (Owen and Dahlin, 2005). No surface flow occurs until the channel aquifer is saturated (Nord, 1985). These alluvial aquifers are often good sources of water for irrigation and domestic use, whether at the large scale (Owen and Dahlin, 2005; Moyce et al., 2006; Raju et *al.*, 2006; De Leon *et al.*, 2009) or the small or artisanal scale (De Hamer *et al.*, 2008; Harrington *et al.*, 2008; Agyare *et al.*, 2009; Ofosu *et al.*, 2010).

Water resources modelling can provide a holistic approach to the water cycle. A holistic approach allows for the evaluation of water resources availability at catchment, sub-basin or basin scale, including surface water, ground water, and vapour transfers. This is critical to evaluating the water resources implications of livelihood strategies at a nexus of stressed water resource availability. Applying water resources modelling to development problems is thus extremely important, with water research currently biased in favour of environmental rather than developmental analysis and too little research is on adaptation to climate change by developing countries (Van der Zaag *et al.*, 2009).

1.2.   Research Framework

1.2.1.   *Research Hypothesis*

Within the variety of food production interventions proposed for rural areas, it is the general hypothesis of this research that the effects of the upscaling of an intervention can be determined, in terms of the number of households which will benefit and the scale of the downstream impact on water resources. It is suggested that such an assessment can be used to help to prioritise development and intervention options, for example in terms of the greatest benefit for the least downstream impact.

1.2.2.   *Research Questions*

Which water resources strategies offer the best balance of benefit to improving food production for rural households compared to downstream impact on water resources availability, in the semi-arid Mzingwane Catchment?

a)   Which types of land and water use related interventions are needed in Southern African drylands to achieve sustainable development, and in particular to attain the MDGs related to food production? (Chapter 2)

b)   What changes have taken place in rainfall, runoff and temperature within the semi-arid study area over the second half of the twentieth century, and what are the associated risks and implications for food production? (Chapter 3)

c)   How do catchments in the study area respond to rainfall, in terms of runoff generation, vs. interception and evaporation, and what are the implications for food production? (Chapters 4 and 5)

d)   What are the characteristics of different scales of alluvial aquifers and what is their potential for supporting food production? (Chapters 6 and 7)

e) What opportunities and limitations are imposed by water resources upon the upscaling of selected food production interventions in the semi-arid study area? (Chapter 8)

f) How many households in the study area can benefit from selected food production interventions, also in light of expected climate change, and what are the downstream impacts? (Chapter 9)

1.3. Study Area – Mzingwane Catchment, Limpopo Basin

The Limpopo Basin is an important transboundary river basin, stressed by low water resource availability (runoff 13 mm a^{-1}) and high levels of water utilisation in many catchments (Boroto, 2001). The northern Limpopo Basin is also referred to as the Mzingwane Catchment (Figure 1.1)

1.3.1. Climate

The Mzingwane Catchment is a semi-arid area, with rainfall decreasing from north to south: on average around 630 mm a^{-1} at Esigodini, to 560 mm a^{-1} at Filabusi, to 360 mm a^{-1} at Beitbridge (Love et al., 2010b). Rainfall has a unimodal seasonal pattern, controlled by the Inter Tropical Convergence Zone and falling between October/November and March/April (Makarau and Jury, 1997). The movement of the ITCZ from the equator marks the start of the rainy season in the southern hemisphere. In a normal year, it fluctuates half way between Tanzania and Zimbabwe but never moves beyond Limpopo River in the south. The ITCZ moves with the sun, southwards at the start of summer (October/November) and northwards in late summer (March/April) (Twomlow et al., 2006). Because of this, the wet season in the southern parts of Zimbabwe (including the northern part of the Limpopo Basin, also known as the Mzingwane Catchment) starts later and ends sooner than in the northern areas. Furthermore, the northern winds in the convergence are moister than the southern winds, leading to less frequent rainfall in the southern areas than the north, for a given air moisture level.

Rainfall in southern Zimbabwe thus occurs over a limited period of time, and often a large portion of the annual rainfall can fall in a small number of events (De Groen and Savenije, 2006; Twomlow and Bruneau, 2000) with high spatial and temporal variability (Love et al., 2011).

Because of these stresses, life and livelihoods in southern Zimbabwe have revolved around the larger rivers since ancient times and given birth to cultures such as Mapungubwe (Manyanga, 2006).

1.3.2. Hydrological Characteristics

It is estimated that around a quarter of the runoff of the Limpopo Basin is generated in the Mzingwane, see Table 1.1.

Table 1.1. Tributary river basins of the Mzingwane and their mean annual runoff to the Limpopo Basin

| Tributary | Catchment area, km^2 | Mean annual unit runoff, mm | | Coefficient of Variation % | Mean annual unit rainfall, mm |
		Görgens and Boroto, 1997	MCC, 2009	MCC, 2009	Station nearest centre of catchment
Shashe (overall)	18,991	24.33			
Shashani *			3.49	409	140
Simukwe *			2.91	484	145
Thuli *			7.91	383	138
Mzingwane	15,695	22.30	15.70	560	138
Mwenezi	14,759	17.34	14.76	472	140
Bubye	8,140	6.51	8.14	345	153

* Major tributaries of the Shashe river in Zimbabwe.

The Mzingwane and its tributaries are ephemeral rivers, drying up during the dry season, with occasional permanent pools (Minshull, 2008).

1.3.3. Reservoirs

This temporal variability of rainfall means there is a need for inter- and intra-annual storage to guarantee water supplies for domestic use and for agriculture. A large number of reservoirs have been constructed to store water from the rainy season and from high rainfall years (Mavimavi, 2004), including large reservoirs to supply cities, mines and commercial irrigation and small dams supplying livestock and smallholder irrigation schemes.

Table 1.2. Selected information on major dams in the northern Limpopo Basin, operated by the government of Zimbabwe. Source: ZINWA database, interviews with water bailiffs. For locations, see Figure 1.1.

Dam	River	Full storage capacity (10^6 m^3)	Year of construction
Mzingwane	Mzingwane	42	1962
Inyankuni	Inyankuni	75	1963
Lower Ncema	Ncema	17	1964
Silalabuhwa	Insiza	23	1966
Upper Ncema	Ncema	45	1973
Insiza (Mayfair)	Insiza	173	1973
Manyuchi	Mwenezi	303	1988
Zhovhe	Mzingwane	136	1995
Glassblock	Mzingwane	14	planned
Oakley Block	Mzingwane	41	planned

1.3.4. Geology and Soils

Geologically, most of the catchment is underlain by the Zimbabwe Craton: mafic greenstone, Shamvaian clastics and Archaean granitoid terrain. The south is underlain by Limpopo Belt Archaean gneisses, and the south-west and far south-east by Karoo and Jurassic basalts, intrusives and sediments, see Figure 1.2.

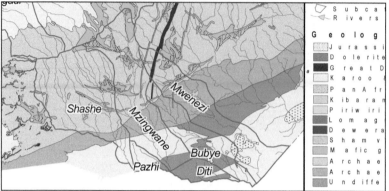

Figure 1.2 Geology of the Mzingwane Catchment, after Ashton *et al.*, 2001.

The soils of Zimbabwe may be broadly categorized into two main groups: (a) old soils formed on deeply weathered parent materials, influenced by earlier erosional surfaces, and (b) relatively young soils, formed on the more recent erosional surfaces, or on alluvial deposits. Deeply weathered ancient soils occur mainly on the plateaus (Highveld), and in some protected areas of the Escarpment zone. These soils have formed over long periods on the weathering mantle or saprolite, and have developed under warm and humid climatic conditions needed for intense chemical weathering. Younger and less weathered soils characterize the denuded hills and mountain ridges and Middle and the Lowveld, where recent and sub-recent erosion has removed any deeply weathered soils. Recent and sub-recent climatic conditions have not been conducive to strong weathering and new formation of saprolite in the eroded areas (Nyamapfene, 1991).

Using soil or vegetation data from a large scale database can be highly problematic, as land use and soil type can vary on the kilometre scale (Vachaud and Chen, 2002). Unfortunately, soil type data in Zimbabwe is available only in at 1 : 1,000,000 scale (DRSS, 1979) and 1 : 250,000 for the communal lands of Zimbabwe, which cover 42 % of the country (Anderson *et al.*, 1993). Typical soil profiles are classified according to the Zimbabwe soil classification system (Nyamapfene, 1991), and have been correlated with the Legend of the Soil Map of the World (FAO, 1988) and Soil Taxonomy (Thompson and Purves, 1978). For the remainder of Zimbabwe, it is necessary to rely on an older 1:1 million scale map (DRSS, 1979). Although there is large uncertainty with respect to the accuracy of the information for some areas, this map is generally widely used to provide soils information for the country.

1.3.5. Hydrogeology

The granite and gneiss backbone of Zimbabwe is a secondary aquifer, with the main components being the weathered regolith and in fractures on the bedrock (Owen, 2000). In such rocks, fracture porosity contributes to higher permeability and transmissivity. Groundwater resource development is possible at locations where favourable lithology, structural features and weathering coincide to form zones of higher transmissivity. Fractures (such as faults, joints, veins), dykes and zones of weathering are targets for the exploitation of groundwater in crystalline rocks (Owen *et al.*, 2007). In the Mzingwane Catchment, granites and gneisses make up more than two-thirds of the bedrock. Their groundwater potential has yet to be ascertained, but preliminary investigation suggest that structural controls on groundwater occurrence in the area are similar to those postulated for elsewhere in Zimbabwe (Basima Busane *et al.*, 2005).

Alluvial groundwater is an especially important resource in arid and semi-arid areas given its sub-annual recharge – much faster than deep groundwater – and lower exposure to evaporation, compared to surface storage (Love *et al.*, 2010c ; Olufayo *et al.*, 2010; Otieno, *et al.*, 2011), and it can be developed for agricultural water supply more easily and cheaply than deep groundwater (Herbet, 1998; Mansell and Hussey, 2005). Alluvial groundwater makes up an important part of the water balance of the Limpopo river and its tributaries (Boroto and Görgens, 2003; Love *et al.*, 2010c). Groundwater is thus an attractive water source in the northern Limpopo Basin (Love *et al.*, 2007; 2010c).

1.3.6. Land Use and Land Cover

The Mzingwane Catchment is covered mainly by a mixture of croplands, pastureland and woodland, see Figure 1.3.

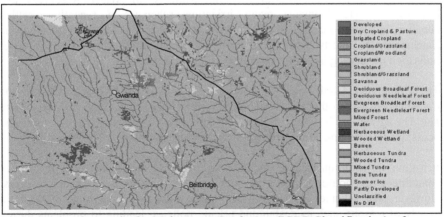

Figure 1.3 Land cover in the Mzingwane Catchment (IGBP Classification), after Hearn *et al.*, 2001. The thick black line marks the northern boundary of the Mzingwane Catchment.

Land use in the catchment is mainly communal lands (smallholder farming) and commercial/resettlement lands (medium to large scale farming), see Figure 1.4.

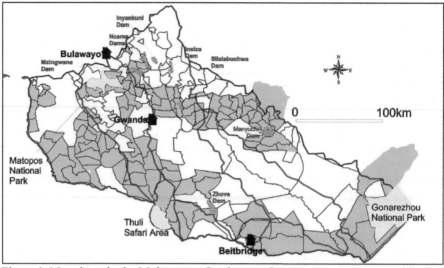

Figure 1.4 Land use in the Mzingwane Catchment. Green areas are communal lands, white areas commercial/resettlement lands and yellow designates national parks estate. Blue lines denote watersheds between the subcatchments. Derived from Surveyor General of Zimbabwe and Forestry Commission (1996) and ICRISAT database.

Commercial agriculture in the north includes cattle ranching and irrigated agriculture, with crops such as maize, wheat and vegetables being grown. In the south, commercial agriculture is mostly ranching, with some large irrigated estates growing citrus or sugar cane. Commercial agriculture may be on privately owned land, ranging in size from 400 ha upwards, or under model A2 resettlement commercial farming land, ranging in size from 200 to 400 ha, depending on natural region (Moyo, 2004). According to the Ministry of Water Resources and Infrastructure Development, many irrigation schemes in recently resettled commercial land throughout Zimbabwe, including within the Mzingwane Catchment, are poorly maintained and require rehabilitation to allow for better utilisation of existing irrigation potential (Mutezo, 2005).

The Communal Lands are where most of the population of the Mzingwane Catchment lives and they are smallholder farmers, holding tenure under customary law. The agriculture is mainly dryland farming, with the major crop being maize and occasionally sorghum, combined with livestock husbandry.

Irrigation in communal lands includes irrigation schemes, managed by farmer committees, and household level irrigation – the latter mostly for vegetable gardens. However, such schemes also tend to over-apply water (Senzanje et al., 2003), for a variety of reasons such as design and pricing policies, leading to problems during drought years. Poor water management on such schemes also leads to reduced yields (Samakande et al., 2004). Furthermore, access to irrigation water for the smallholder farmer remains limited. Despite this, many dams developed for irrigation (and other) purposes are heavily underutilised, with irrigation from Mchabezi and Zhove dams being initiated more than five years after their construction.

There has been widespread introduction of low cost drip irrigation kits in the communal lands (Chigerwe *et al.*, 2004; Maisiri *et al.*, 2005). A recent study showed that only 2 % of the beneficiaries had used the kit to produce the expected 5 harvests over 2 years, owing to problems related to water shortage and also pests and diseases. About 51 % of the respondents had produced at least 3 harvests and 86 % produced at least 2 harvests. Conflicts between beneficiaries and water point committees or other water users developed in some areas especially during the dry season (Moyo *et al.*, 2006).

1.3.7. *Water Quality and Pollution*

Ambient river water quality in the upstream tributaries is generally satisfactory. Water in some river reaches show high levels of metals such as cadmium, iron and zinc. This is partly an ambient condition and partly due to pollution from mining (Love *et al.*, 2006b). The latter is mainly confined to the vicinity of the mines, and better water quality resumes downstream, as has been seen elsewhere in Zimbabwe (Lupankwa *et al.*, 2004; Ravengai *et al.*, 2004, 2005*a*, 2005*b*). Gold panning is also a major problem, causing siltation of rivers and mercury pollution, including in the study area (Shoko and Love, 2005).

Many of the alluvial aquifers in the downstream catchments, especially smaller aquifers and those on river bank flood plains, are characterised by high levels of sodium and chloride. This is an ambient condition, related to the geology of the aquifers, and threatens irrigated agriculture with equipment or crop failure mining (Love *et al.*, 2006b).

Results of a limited study of variations in groundwater chemistry from the Mwenezi and Gwanda areas suggest salinity and turbidity problems (Hoko, 2005). A stable isotope study carried out in the adjacent Save Basin, suggests that salinity could be related to localised evaporite deposits, rather than larger scale characteristics of the regional flow system (Sunguro, 2001).

1.3.8. *Institutional Arrangements for Catchment Management*

One of the major initiatives in water management in southern Africa, developed through national water reforms since 1990, has been decentralisation of management from central government to some form of localised water authority, with varying degrees of stakeholder participation and control (Jaspers, 2003). Water sector reform in Zimbabwe has been implemented via the creation of two parallel structures: a parastatal (the Zimbabwe National Water Authority, ZINWA) and stakeholder councils (Catchment Councils and their national forum). There are seven water management areas which are termed "Catchments". The seven Catchments are based on hydrological boundaries, but not catchments in a hydrological sense: four are portions of the Zambezi Basin defined by major tributaries: Gwayi, Sanyati, Manyame and Mazowe), two are portions of the Save Basin (Runde and Save) and one – the river basin studied in this thesis – is a portion of the Limpopo Basin (Mzingwane).

ZINWA consists of a head office in the capital city and offices in each Catchment. The head office, and the authority, falls under a Chief Executive Officer (CEO), to whom report heads of departments of the authority, as well as each of the seven catchments, which are headed by a Catchment Manager. The stakeholder councils include a forum at national level, Catchment Councils and Subcatchment Councils. The Subcatchment Councils comprise elected or nominated stakeholder representatives. Government officials of departments with responsibilities pertinent to water, such as the Department of Natural Resources and the Department of Agricultural Research and Extension, are non-voting members of Subcatchment Councils. Thus, the Subcatchment Councils can provide a platform for stakeholder engagement and participation, even if they do not hold decision-making powers in some areas, such as catchment conservation (Zwane et al., 2006).

The Catchment Councils comprise representatives elected by the Subcatchment Councils within the Catchment and the ZINWA Board includes representatives elected by the Catchment Councils and other stakeholders appointed by the minister responsible for water (Latham, 2002). A significant problem in stakeholder representation is that many users are not recognised directly as users, but rather represent political authorities: in the rural areas communal farmers are represented by the Rural District Councils and urban residents are represented by the urban councils (Manzungu and Mabiza, 2004).

These structures are parallel without direct reporting relationships between them: for example, although the Catchment Council and the Catchment Manager are responsible for the same geographical (and hydrological) area, the Catchment Manager reports to the CEO, and only consults the Catchment Council. According to the law, the Catchment Manager is responsible for water resources management, administration and the control of water utilities, whereas the Catchment Council is responsible for water allocations and planning. In practice there is much overlap: Catchment Managers can allocate blue water use permits and some are leading or controlling the planning process. The result is that whilst many issues are discussed at the stakeholder councils, power remains with the local offices (Catchment Managers) of the national authority (Nare et al., 2006). Because of this, Catchment Councils have come to be viewed by some stakeholders as an extension of ZINWA and thus of government, rather than authorities in their own right (Sithole, 2001).

In many of the Catchment Councils, control remains with the large, powerful users of blue water: city councils, large mines and large-scale commercial farmers. The Mzingwane Catchment Council is dominated by large-scale commercial farmers, cities and large mines (Nare et al., 2006). Such powerful users also generally participate far more in water management and planning than other users do (van der Zaag, 2005). This trend can be related to two factors: such large users had previous experience in the (now dissolved) River Boards and the issues covered tend to be on permits and levies, which apply mainly to the larger users (Latham, 2002).

In terms of local governance, most of the Mzingwane Catchment lies within Matabeleland South Province, under a Provincial Governor in Gwanda. Some parts of the Mwenezi Subcatchment fall into Midlands Province and Masvingo Province. Below the provinces are the districts, run by District Administrators (DAs) (officials who are part of the Ministry of Local Government and National Housing) and Rural District Councils (RDCs), which have elected councillors and chairpersons. The local government and water management boundaries do not coincide. Most governmental functions in Zimbabwe are carried out by national government departments, even at the local level in rural areas. Thus, for example, health officers at district level report to a provincial health officer who reports to the permanent secretary for health. All three officials are part of the national Ministry of Health and Child Welfare, and not any district (local government) or provincial government department.

In practice, governance activities in rural areas tend to operate by consultation between the RDC, DAs and local officials of national government departments. This is not required by the constitution or law, but tends to be the *modus operandi*. The key officials are the DA and the RDC CEO, since they manage district officials and liase with and coordinate with district / local level officials of national government departments (ministries) and other national structures (eg ZINWA). It should be noted that there is a high turnover of governmental officials at all levels, due to generally uncompetitive salaries, and this makes it difficult to retain good staff and, consequently, to realise long-term planning, management and implementation of policies in the water sector.

Relevant decision-making can be conceptualised at six scales: household, village, ward, Subcatchment water management authority, Catchment water management authority or Province and National Government. Above this lies the Limpopo Basin and the transboundary Limpopo Basin Commission; that scale is not considered in this thesis.

At household of family level, decisions are taken on field management, crop management (crop selection, cropping programme, irrigation method if any), tillage method (sometimes), water use and livestock management.

At village level, including the headman (*sabhuku*) and the Village Development Committee (VIDCO), decisions are taken on land allocation (in communal lands), cropping programme (sometimes), irrigation method (sometimes), tillage method (sometimes). Community groups of farmers develop and labour is shared, hired and sold.

At ward level, including the chief, local councillor and the Ward Development Committee (WADCO), decisions are taken on land allocation (in communal lands) and crop management (sometimes). Training is provided by agricultural extension staff, e.g. in tillage methods.

At District or Subcatchment level, including the Rural District Council, Subcatchment Council and Member of Parliament, decisions are made on land allocation (in resettlement lands) and water allocations for large users. Training is provided by natural resources staff, e.g. in soil conservation.

At Catchment water management area or Provincial level, including the Governor, Provincial Development Committee and Catchment Council, decisions are made on water conflict resolution, siting of dams and irrigation schemes, land allocation (in resettlement lands) tillage and irrigation methods to promote.

At National Government level, decisions are made on policies, legislation, water resources development, food security, food distribution and marketing, macro-economics.

1.3.9.   Field Study Sites

Five field sites were instrumented during this study (Table 1.5; see Figure 1.1 for locations). Further information on the study sites is given in the following chapters (sections 4.3.1, 5.3.1, 6.3.1 and 7.2.3).

Table 1.5. Field sites instrumented during this study

Catchment	*Discharge stations*	*Climate stations*	*Hydrogeological stations*
Lower Mzingwane	Bridge and limnigraph	16 rain gauges Class A evaporation pan	4 piezometer arrays
Mnyabezi 27	Dam and limnigraph	7 rain gauges Class A evaporation pan	4 piezometer arrays
Mushawe	Bridge and limnigraph	17 rain gauges Class A evaporation pan	4 piezometer arrays
Upper Bengu	Dam and limnigraph	8 rain gauges Class A evaporation pan	---
Zhulube	Composite gauge (V-notch and broad crest)	14 rain gauges Class A evaporation pan	---

1.4.   Outline of the Thesis

This thesis consists of seven substantive chapters, as well as an introductory and a synthesis chapter. Chapter 2 (based on Love *et al.*, 2006) gives a literature review on the land, water and livelihood strategies in the context of the Millennium Development food security goals. In chapter 3 (based on Love *et al.*, 2010b), changes in the rainfall and discharge patterns in the study area are analysed statistically, in the context of climate change.

Chapter 4 (based on Love *et al.*, 2010a) describes the development of a rainfall-interception-evaporation-runoff model in the study area and chapter 5 (based om Love *et al.*, 2011) deals with its regionalisation. Two studies on the modelling of alluvial groundwater and groundwater – surface water relationships are presented: at meso-catchment scale (chapter 6, based on Love *et al.,* submitted) and river sub-basin scale (chapter 7, based on Love *et al.,* 2010c).

In chapter 8, the rainfall-interception-evaporation-runoff model and the groundwater – surface water model are coupled, to provide a holistic model for the evaluation of the implications of livelihood strategies and climate change for water resources management in the study area.

Finally, in chapter 9, a synthesis is made of the scenarios modelled in chapters 6, 7 and 8 and the major findings of this research are summarised. Based on these results, development recommendations are made, and future research directions are discussed.

2. Implementing the millennium development food security goals - challenges of the southern African context[*]

2.1. Abstract

The Millennium Development Goals' target to halve the proportion of people who suffer from hunger is extremely important in southern Africa, where food security has become increasingly problematic over the last 20 years. One "quick-win" proposal is replenishment of soil nutrients for smallholder farmers, through free or subsidised chemical fertilisers. Other proposals include appropriate irrigation technology, improved inputs and interventions targeted at women.

Analysis of over 10 years of agro-hydrological and agro-economic studies from southern African show that a different approach is required to interventions proposed. There are sustainability problems with free chemical fertiliser due to transport costs and ancillary costs. Furthermore, recent studies in Zimbabwe and Mozambique show that significant increases in yield can only be obtained when soil fertility management is combined with good crop husbandry, e.g. timely planting and weeding. Ongoing replenishment of fertility would be dependent on a continued free or subsidised fertilizer supply, and transport system. Increasing access to irrigation will help, but is not the only solution and cannot reach even a majority of farmers. It has been determined that short dryspells are often the major cause of low yields in sub-Saharan Africa. Soil-water conservation approaches, e.g. winter weeding and conservation tillage, can reduce risk and increase yield.

The following specific recommendations are made for urgent interventions to contribute sustainably to food security in southern Africa: (i) To increases access to fertiliser, consider development of strong input markets at end-user level. (ii) Intensification of technology transfer, focusing on capacity building for transfer of existing technologies and much closer collaboration between state and NGO sectors, agronomists and water engineers. (iii) Increasing the uptake of soil-water conservation methods, including conservation tillage and weeding, and supplementary irrigation to minimize adverse effects of dryspells, through investments in farmer training. (iv) Linking crop development strategies to livestock development practices and strategies. (v) Developing non-agro-based livelihood strategies in marginal lands.

2.2. The millennium development project and proposals for implementing the goals

The first Millennium development goal (MDG) aims to eradicate extreme poverty and hunger. Target 2 of Goal 1 is to halve, between 1990 and 2015, the proportion of people who suffer from hunger. In order to achieve this, the Millennium Project

---

[*] Based on: Love, D.; Twomlow, S.; Mupangwa, W.; van der Zaag, P.; Gumbo, B. 2006a. Implementing the millennium development food security goals - challenges of the southern African context. *Physics and Chemistry of the Earth*, 31, 731-737.

proposes some initial, urgent measures and some longer term proposals. The interventions to be started immediately for quick results are referred to as the "quick-wins". In addressing food security (target 2), the Millennium Project proposes, as a quick-win, "a massive replenishment of soil nutrients for smallholder farmers on lands with nutrient depleted soils, through free or subsidised distribution of chemical fertilisers and agroforestry, no later than the end of 2006" (UN Millennium Project, 2005a). The emphasis on smallholder farmers is based on an assessment that of the 850 million people experiencing serious and chronic hunger, approximately 50 % are smallholder farmers (FAO, 2004). For this reason, the Millennium Project recommendations on rural development and food security focus on improving the production and livelihoods of smallholder farmers. The Project's major concerns for agriculture are soil fertility, water resources management, access to improved varieties of crops and livestock and agricultural extension services (UN Millennium Project, 2005b).

Detailed intervention proposals from the Millennium Project include:
1. Investments in soil health: application of mineral fertiliser, agroforestry (use of trees to replenish soil fertility), erosion control, return of crop residues to the soil.
2. Appropriate irrigation technology, which the Millennium Project confusingly refers to as "small-scale water management": drip irrigation, wells, appropriate pumps and so on.
3. Improved inputs: improved varieties of crops, pastures, livestock, trees and fish.
4. Farm diversification to include higher value livestock and vegetables, once farm/household level food security has been achieved.
5. Extension services require strengthening, especially at village level, with up-to date agricultural knowledge and participatory approaches to farmer training.
6. Agricultural research requires investment at national level.
7. Interventions must be targeted to ensure empowerment of women farmers, including women extension officer recruitment, and promotion of women's property/tenure rights.

The MDG target to halve the proportion of people who suffer from hunger is extremely important in southern Africa, where food security has become increasingly problematic in the last half-century. Recurrent droughts and inconsistent national and regional policies fail to provide adequate support to large scale and small scale producers, leading to a net decline in agricultural production in the region. The poor performance of agriculture in Africa has been well documented. Per capita food production has not matched the growing population over the last 40 years (FAO, 2004). 180 million of the world's 800 million food insecure people live in sub-Saharan Africa, surviving on less than 1 US dollar per day.

Therefore, implementation of the food security target of the millennium development goals is critical to the maintenance and improvement of livelihoods in southern Africa. However, as this paper will show, developing solutions at a global level does not always capture properly the needs and challenges of regional contexts. This is particularly important in considering the "quick-win" strategies proposed by the Millennium Project.

In this study, a review of the challenges to crop production adaptations to interventions proposed by the Millennium Project to work towards the MDGs are discussed. The specific hydrological and agro-hydrological conditions of southern Africa are particularly important. Accordingly, some modification of the approach is called for.

2.3.   Challenges to crop production in southern Africa

2.3.1.   Context

Africa's population is expected to reach 1.2 billion by the year 2020, double the 1995 figures. It is estimated that at least 25 % of the population will be undernourished and living in the dryland areas of sub-Saharan Africa, which already accommodate 70 % of the world's poorest communities (Ryan and Spencer, 2001). Sub-Saharan Africa is the only region in the world where average food production per capita has been declining over the last 40 years. Yet agriculture continues to be the dominant economic activity, accounting for 70% of total employment, 40% of total exports and 34% of the GDP.

Despite the technological advances in agricultural production in recent years, poverty, food insecurity and malnutrition still remain major challenges in sub-Saharan Africa (Sanchez and Swaminathan, 2005), yet agriculture is the main economic activity in the region. Added to this are natural and human-induced conflicts, and low investment in long term capacity building and research. It has been argued that short-term land tenure arrangements may not encourage long-term investment on the land, especially where the tenure system is unclear or dynamic (e.g. Adams et al., 1999; Gebremedhin et al., 2003; Smith, 2004). That said, many improvements, especially in soil/water conservation, are compatible with a short-term investment perspective (Gebremedhin et al., 2003) and can be made on an annual rather than permanent basis, under any tenure system.

The most vulnerable groups are smallholder land users in the arid and semi arid lands (Freeman et al., 2002). Smallholder farmers make up more than half the population at risk from hunger, and in southern Africa cultivate on poor soils and unreliable rainfall is the main source of water (Twomlow and Bruneau, 2000). Crop yields are low and failures are frequent (Scoones, 1996), with more than one million people in the Limpopo basin depending on food aid in 2003 (Love et al., 2004). In this context, agricultural production in southern Africa faces a number of challenges, each of which has implications for the Millennium Project's intervention proposals.

In figure 2.1, some of the major challenges are summarised.

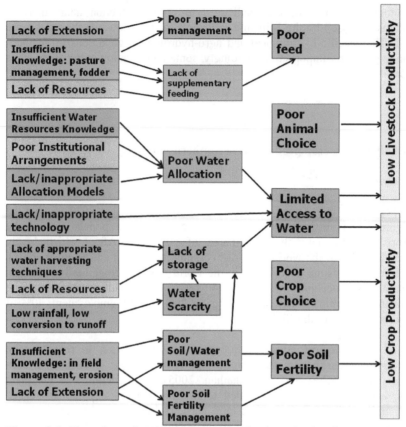

Figure 2.1. Flow chart of major challenges to food production in the semi-arid Limpopo Basin. Adapted from Nyabeze and Love (2007): blue boxes are water-related challenges, green boxes are soil/crop/livestock –related challenges and orange boxes are economics.

2.3.2. *Water*

Smallholder agriculture in southern Africa is largely rainfed, and thus risky, with recurrent droughts and dryspells. Potential evaporation exceeds rainfall for more than 6 months of the year. Rainfall is seasonal and highly variable. Annual rainfall for a single site can vary by up to 1000 mm a^{-1} from year to year (Twomlow and Bruneau, 2000) - although a drought year may record less than 250 mm a^{-1}, such as the 2004-2005 season in southern Zimbabwe and Mozambique. By the end of the dry season, i.e. just before planting, the top 0.3 m of the soil horizon frequently holds negligible water content (Twomlow and Bruneau, 2000). Furthermore, a number of climate change models predict that southern Africa shall experience significantly reduced precipitation and runoff over the next fifty years (Arnell, 2003; Moyo *et al.*, 2005). These trends will result in further food shortages (Cane *et al.*, 1994; Du Toit *et al.*, 2001).

Frequently, the water yield from the developed surface water resource falls short of the demand, deficits being more evident during droughts (Nyabeze, 2004). Much of sub-Saharan Africa has, as a result of low conversion of rainfall to runoff, a precarious balance between available water resources and water demand (Mazvimavi, 2003). Furthermore, demand from more powerful sectors, such as urban areas and commercial farming, is rising. It has been suggested that the expansion of the SADC regional economy and especially the South African economy will necessitate water re-allocation from agriculture to urban areas and industry (Taigbenu et al., 2005). Even with such limitations, smallholder farmers have little access to blue water for irrigation (Love et al., 2004). This is partly due to a lack of investment targeting such developments, but also since, in some areas, there is a lack of suitable sites to establish further dams. Some catchments have been over-committed for quite some time (e.g. Kabel, 1984), meaning that additional surface water development is not possible, and others offer few sites where a possible river stretch for impoundment is coupled with suitable soils. Furthermore, most southern African governments face a chronic shortage of funding for capital investments such as dams.

2.3.3. Soils

Smallholder farmers generally farm on poor quality sandy or sandy loam soils (Twomlow and Bruneau, 2000). Frequently, such soils are infertile, deficient in nitrates, phosphates and sulphur (Burt et al, 2001). This is especially the case in those countries were colonial imbalances in land distribution have not been adequately adjusted. Even in countries where such attempts have been made, smallholder farmers are generally confined to the poorer soils. Rising populations have placed an increasing burden on the soils farmed by smallholder farmers: in the last half-century in southern Africa the populations have increased dramatically, but the availability of land to smallholder farmers has not increased significantly in most countries. Thus a higher population is depending on the same area of semi-fertile land (Moyo, 2004).

Fertiliser usage is low in southern Africa, compared for example to India: 28 kg/ha compared to 62 kg/ha (Twomlow et al., 1999). Due to lack of investment in soil fertility amendments to cope with declining yields, farmers have extensified their cropping area. As soil fertility declined, farmers have maintained household food requirements by increasing the cropped area to compensate for lower yields.

2.3.4. Economics and Livelihood Strategies

The question of livelihood strategies is not addressed adequately in the MDGs. Rural families livelihood strategies are not necessarily geared directly to food production, many household have diverse coping strategies to ensure some degree of sustainable livelihoods. Livestock (which are not generally a food source but a strategic financial asset) are valued ahead of crop production. For example, farmers in many southern African countries prefer to feed crop residues to their cattle than to mulch or plough in the residues. This is despite the well-proven benefits in terms of soil fertility improvement, and also increased infiltration from precipitation (Bennie and Hensley, 2001).

Studies have shown that poverty in southern Africa presents a greater driver to food insecurity than actual food production. It has been suggested that access to food can be a greater problem for households than absolute food availability (Misselhorn,

2005). This can be partly due to few smallholder farmers subsisting entirely on their own production, and must still sell some produce to buy some of their food. Since it is well established that world food production is sufficient, but distribution inadequate (e.g. Myers, 1985), it is not logical to resolve food security problems by focussing solely on increasing production.

2.4. Adapting and operationalising the MDGs for food security in southern Africa

2.4.1. Fertiliser

The quick-win proposal on mineral fertiliser is especially difficult. Accelerating and increasing fertiliser use does not deal with the underlying causes of declining soil fertility in southern Africa. Thus, while an urgent and massive investment in mineral fertiliser will have a short-term impact; it is unlikely to make any impact on food security in the medium term. It should be realised that a massive investment in mineral fertiliser does not address long term food security problems: there is no guarantee that such an investment would translate into a sustainable improvement in soil health. Ongoing replenishment of fertility would be dependent on a continued, guaranteed free or subsidised supply, transport system and so on.

The low rate of fertiliser application in southern Africa reflects the much higher end user price of fertiliser in Africa. This is often due to transport and import costs, especially in landlocked countries. Smallholder farmers are generally living distant from the major urban centres of their countries, increasing transport costs for them. This leads to a vicious circle at the smallholder level where current recommendations are inappropriate: few farmers have had positive experiences of heavy fertiliser use, thus there is little demand, and thus rural tradestores do not stock much variety of fertiliser. Occasional free fertiliser handouts, by government or NGOs, tend to destabilise the small local markets, in countries where fertiliser production is done by private companies. Such interventions thus provide short-term benefits but medium-term problems. These ancillary costs will remain to limit smallholder farmers in southern Africa, even if the original procurement of fertiliser is free or heavily subsidised internationally.

Fertiliser is insufficient in itself to produce a major change in the food security situation. Major increases in yield can only be obtained when soil fertility management is combined with soil-water conservation practices such as timely planting and weeding, thus reducing periods of potential moisture stress/competition (Twomlow et al., 1999). For these reasons, massive investment in mineral fertiliser will not produce a "quick win" unless it is integrated with transport and other interventions.

2.4.2. Water Resource and Wate Use Strategies

There is an important difference between:
- Overall physical water resource availability (controlled mainly by climatic and hydrological processes), and

- Water resource availability at household scale, which is controlled by the above physical factors, and by access.

Irrigation schemes have a role to play in imporving food production, and smallholder irrigation schemes can become a critical common property resource (Samakande et al., 2004). However, such schemes also tend to over-apply water (Senzanje et al., 2003), for a variety of reasons such as design and pricing policies. Water allocation between farmers is not always equitable or efficient, especially during drought years (Munamati et al., 2005). Poor water management on such schemes also leads to reduced yields (Samakande et al., 2004). Furthermore, access to irrigation water for the smallholder farmer remains limited. Thus while there is considerable scope of expansion of irrigation in Sub-Saharan Africa, it is also imperative to look at improving land and water management in the schemes (De Fraiture and Wichelns, 2010).

The Millennium Project recognises the limitations of large scale irrigation for food security in Africa and urges the use of appropriate irrigation technology such as low-cost drip kits. This can allow for greater access to irrigation, since the water and energy requirements are lower than conventional irrigation (Chigerwe et al., 2004). It can also allow for irrigation at smallholder household level, rather than smallholder scheme level. Drip technology on its own tends to improve water use efficiency, but sometimes does not increase yield (Maisiri et al., 2005). A recent study showed that only 2 % of the beneficiaries had used the kit to produce the expected 5 harvests over 2 years, owing to problems related to water shortage and also pests and diseases. Conflicts between beneficiaries and water point committees or other water users developed in some areas especially during the dry season (Moyo et al., 2006). Thus the implementation of small-scale irrigation has to be done in a manner to address these challenges.

As an alternative to permanent irrigation, yields can be improved by short-term supplementary irrigation during dryspells (Savenije, 1998, 1999; Nyamudeza, 1999; Falkenmark and Rockström, 2003; Rockström et al., 2003). As an alternative to dams, use can be made of rainwater harvesting technologies (Motsi et al., 2004) or accessing of alluvial aquifers (Dahlin and Owen, 1998; Moyce et al., 2005).

However, small-scale irrigation as promoted by the millennium project (e.g. Sanchez and Swaminathan, 2005) is not the only solution. Irrigation schemes can never benefit more than a small minority of farmers in a district, since there may not be sufficient water resources, and an investment of an order of magnitude way beyond that proposed by the millennium project would be required. To improve yield and livelihoods for a larger proportion of smallholder farmers, a farm-scale approach of integrated soil and water management is needed for dryspell and drought mitigation. Short dryspells - not necessarily water scarcity measured at an annual scale - are often the major cause of low yields and loss of food security (Rockström et al., 2003). Improvements can be achieved through agricultural interventions in dryland farming (Twomlow et al., 1999). Some in field water management technologies, such as tied ridges, have proven effective at reducing runoff, thus increasing infiltration and water access to the crops (Heinrich, 2001; Motsi et al., 2004) - although uptake is variable (Twomlow et al., 1999). Conservation tillage approaches, especially where they are adapted for use with animal traction, have proven very effective in increasing water availability to crops and decreasing land degradation (Kaumbutho and Mwenya, 2000; Mupangwa *et al.*, 2006; Twomlow *et al.*, 1999).

Effective weeding also increases both in-field water availability and crop yield. Biomass can be increased by a factor of four and water use efficiency (kg/mm) by a factor of five, for well weeded maize compared to an unweeded control (Twomlow *et al.*, 1999). Weed growth influences negatively the availability of water in the soil profile, especially during dryspells (Twomlow and Bruneau, 2000). It is not only rainfed farming, but also smallholder irrigation schemes that can benefit from improved soil-water management (Samakande *et al.*, 2004).

Often, the problem is not physical scarcity of water, but rather inadequate access to water resources at household scale, or poor management of scarce water resources at farm scale. This can come about due to lack of integrated management approaches (linking crops, soil, water, climate), human and financial capacities and weak institutional arrangements. The new water governance in the Limpopo Basin countries enshrines principles of equity and user participation in decision-making (Love *et al*, 2004). Catchment institutions provide a possibility for progressive and sustainable, stakeholder-based water management (Manzungu and Mabiza, 2004; Savenije and van der Zaag, 2000; van der Zaag and Savenije, 2000). However, penetration remains problematic: smallholder farmers tend to have very little voice, while large water users dominate (Love *et al*, 2004; Nare *et al.*, 2006; van der Zaag, 2005).

2.4.3.   Other Interventions

Access to improved varieties for smallholder farmers is often limited, and faces many of the same limiting factors as fertiliser distribution. Furthermore, many classic "green revolution" improved varieties are not drought resistant – and produce the expected high yields only when free of water stress. Finally, if crop management is not improved at the same time as new varieties are adopted, the farmer will never see the real yield potential.

A significant level of farm diversification already exists in southern Africa, often focusing on livestock, as has been discussed under livelihood strategies above. However, additional farm diversification requires developments of a market for the new produce, and a transport strategy to ensure that the value added in the new produce is not lost to transporters.

Women farmers actually spend substantial amounts of their time and labour on activities not directly connected with agricultural production. Thus developing improved efficiencies of transport and cooking fuels, and improving ease of access to water for domestic purposes would release a large proportion of the woman farmer's labour for her to invest on her farm.

In many countries in southern Africa, what is needed is not retraining of extension staff – much of the most recent knowledge is well known – but resources for expansion of service work and improvement of the conditions of staff to ensure capacity retention.

2.5.   The way forward

The agricultural, climatic, sociocultural and economic context of southern Africa presents challenges that require modifications and alternatives to the interventions proposed by the Millennium Project to work towards the MDGs. Specific recommendations are now made for urgent interventions in agriculture in southern Africa, in order to contribute to food security, as an alternative to the Millennium Project's fertiliser subsidy proposal. Firstly, interventions in inputs (seed and fertiliser) should not be simple subsidies, but should be sustainable investments targeted at either end-user level or at the transport and marketing system. Upgrading the distribution system for inputs will have a longer term and more sustainable impact than free distribution of fertiliser for one or two seasons. The constraint posed by transport is clear, since it has been cited as a major constraint to application of manure, which is readily available (Rohrbach, 2001). Developing the market for inputs, especially for sale of small quantities of inputs, will allow a contribution from the small-scale enterprise sector (Sanders, 2002), and facilitate the development of a sustainable fertiliser distribution network in smallholder farming areas.

Increasing the application and uptake of soil-water conservation methods, including conservation tillage and weeding, is needed in dryland farming and in supplementary irrigation. The interventions should be carried out mainly through intensive investments in farmer training, by collaborating government services (both agricultural extension and state water authorities) and NGOs. The approach must be participatory, to determine what techniques smallholder farmers have themselves found effective in the drier years, and involve farmers in the moitoring and evaluation (Gupta, 2002). This has to be the priority area in extension work, as water management is often the limiting factor in yield and this proposed intervention is more practical and affordable than a massive expansion of irrigation. This intervention is not a "new story", but is deserving of a higher priority than has been indicated by the Millennium Project, since it can benefit a greater number of food insecure people than the proposed fertiliser subsidy.

Alternative livelihood strategies need developing and promoting, especially in the semi-arid to arid areas. One adaptation is to link crop development strategies to livestock development. However, for as long as rural populations increase, and expect to engage in agriculture as their major productive activity, without an increase in land allocation to smallholder farmers, soil fertility and food security will continue to decline, as land is over-utilised (Twomlow and Bruneau, 2000). Accordingly, investment should be made in developing non-agro-based livelihood strategies in marginal lands, including the harvesting of indigenous rangeland products, such as the mopani worm, wildlife management and so on. Given the extent to which poverty, rather than food production, influences food security in southern Africa (Misselhorn, 2005), this could have a major impact.

Water resources planning and evaluation is required before and during implementing water resource interventions, whether these are conventional irrigation schemes, drip kit distribution or supplementary irrigation. It must be ascertained that the water supply assumed for an irrigation project is sufficient, sustainable and acceptable to the community, especially to non-beneficiaries. It is essential that the agriculturalists and the water engineers should engage with each other; preferably at the local government and water management area levels. Proper planning of this kind is likely to decrease the risk of conflicts between water users after implementing the irrigation intervention, and decrease the risk of intervention failure to lack of water supply. It is in this context that this study seeks to make a contribution: this study does not present new evidence to support the likelihood of success of interventions in terms of increasing food production, but rather analyses the hydrological aspects of interventions which have been identified from literature and current practice by governments and other development agents. This is done to determine the constraints imposed by water resources availability on specific interventions and their downstream impact.

3. Changing hydroclimatic and discharge patterns in the northern Limpopo Basin, Zimbabwe[*]

3.1. Abstract

Changing regional and global trends in climate and discharge, such as global warming-related declines in annual rainfall in south-eastern Africa, are likely to have a strong influence on water resource availability, and to increase livelihood risk. It is therefore important to characterise such trends. Information can be obtained by examining and comparing the rainfall and runoff records at different locations within a basin. In this study, trends in various parameters of temperature (4 stations), rainfall (10 stations) and discharge (16 stations) from the northern part of the Limpopo Basin, Zimbabwe, were statistically analysed, using the Spearman rank test, the Mann-Kendall test and the Pettitt test. It was determined that rainfall and discharge in the study area have undergone a notable decline since 1980, both in terms of total annual water resources (declines in annual rainfall, annual runoff) and in terms of the temporal availability of water (declines in number of rainy days, increases in dry spells, increases in days without flow). Annual rainfall is negatively correlated to an index of the El Niño – Southern Oscillation phenomenon. The main areas of rising risk are an increasing number of dry spells, which is likely to decrease crop yields, and an increasing probability of annual discharge below the long-term average, which could limit blue-water availability. As rainfall continues to decline, it is likely that a multiplier effect will be felt on discharge. Increasing food shortages are a likely consequence of the impact of this declining water resource availability on rain-fed and irrigated agriculture. Declining water resource availability will also further stress urban water supplies, notably those of Zimbabwe's second-largest city of Bulawayo, which depends to a large extent from these water resources and already experiences chronic water shortages.

3.2. Introduction

Rainfall in south-eastern Africa is temporally and spatially intermittent (Unganai and Mason, 2002). Annual rainfall for some sites can vary by up to 1 000 mm^{-1} from year to year (Twomlow and Bruneau, 2000). A drought year may record less than 250 mm a^{-1}, such as the 2004-2005 season in the Limpopo Basin (Love et al., 2006a). Rainfall variability in south-eastern Africa is strongly influenced by the coupled ocean-atmosphere El Niño – Southern Oscillation phenomenon (ENSO) (Trenberth et al., 2007). Positive ENSO anomalies generally result in reduced rainfall in the region and are becoming more common (Makarau and Jury, 1998; Alemaw and Chaoka, 2006). Furthermore, there has been a general decline in rainfall in Southern Africa since 1961 (New et al., 2006), with the period 1986-1995 being the driest decade of the 20th century (Trenberth et al., 2007). General circulation models developed with the Intergovernmental Panel on Climate Change Special Report on Emission Scenarios (IPCC SRES) scenarios suggest that annual

[*] Based on: Love, D.; Uhlenbrook, S.; Twomlow, S.; van der Zaag, P. 2010b. Changing rainfall and discharge patterns in the northern Limpopo Basin, Zimbabwe. *Water SA*, 36, 335-350.

rainfall in south-eastern Africa will decline further under the impact of global warming (Christensen et al., 2007). This is expected to be between 10% and 20% below the 19001970 averages by 2050 (Milly et al., 2008) or by up to 10% below the 1980-1999 averages by 2099 (Christensen et al., 2007).

Declines in rainfall, as discussed above, may translate into more than proportional declines in discharge due to non-linear processes, including, for example, interception thresholds. By the years 2041-2060 in south-eastern Africa, runoff is expected to decline by between 10% and 40% compared to 1900-1970 averages (Milly et al., 2005: Fig. 4a).

These changing regional and global trends in climate and discharge are likely to have a strong influence on water resource availability, and increase livelihood risk. It has been shown that household food security in southern Africa is highly vulnerable to climate stress (Archer et al., 2007). Already in much of southern Africa there is a precarious balance between available water resources and water demand as a result of generally low conversion of rainfall to runoff and potential evaporation exceeding rainfall (e.g. Mazvimavi, 2003). Frequently, the water yield from the developed surface water resource falls short of the demand, deficits being more evident during the frequent droughts (e.g. Nyabeze, 2004). Furthermore, some catchments, especially within the Limpopo Basin, are already over-committed (Kabel, 1984; Basson and Rossouw, 2003), leading to water stress: a high ratio of water withdrawal or water use to discharge (Vörösmarty et al., 2000). Changes of this nature constitute a major challenge to water resources management (Milly et al., 2008).

Information on water resource availability, past and present trends and future predictions, can be obtained by examining and comparing the rainfall and runoff records at different locations within a basin. Multi-decadal temporal trends of discharge, rainfall and temperature can be analysed to assess risk to strategic water resource systems (e.g. Hundecha and Bárdossy, 2005; Chen et al., 2007) or to attribute causes of change (e.g. Anderson et al., 2001).

In this study, temperature, rainfall and discharge trends from a series of locations in the northern part of the Limpopo Basin, Zimbabwe, are analysed, in order to determine the:
- Stationarity of the time series and any observed changes on water resource availability
- Causes of any changes observed
- Relationships between rainfall and discharge in the study area and causes thereof
- Implications of any changes

3.3. Methods

3.3.1. Data series

A statistical analysis was carried out of daily temperature, rainfall (Table 3.1) and runoff data (Table 3.2) from several stations (see Figure 1.1 for locations). Stations with a minimum of 30 years continuous data were selected. Data were organised into the Southern African hydrological year of October to September.

All selected discharge stations were upstream of major dams and with limited upstream water users (see Table 3.2 for water use upstream of the discharge stations and Figure 1.1 for the locations of the discharge stations in relation to major dams) and did not require naturalisation.

The time series were visually inspected, along with supporting materials such as the station files. The following exclusions were made for each station, in order to remove unreliable data:
- Where data were missing for 2 months or more, the year was excluded
- Where data were missing for 2 weeks or more during the months of November to April (rainy season), the year was excluded
- Where a note had been made in the station file that readings were unreliable (e.g. due to siltation, security), the year was excluded
- Where the runoff coefficient for a given year was more than 1, the rainfall data were checked for correspondence with the next nearest climate station. If this was unsatisfactory, then the discharge data for that year were excluded.
- Where a daily average flow of more than zero but less than 0.02 m^3/s was recorded for 1 month or more, the year's data for 'number of days of no flow' was excluded as being below the detection level of the gauge. The data selected for analysis, following these exclusions, are shown in Tables 3.1 and 3.2.

The Multivariate ENSO Index (MEI) used was developed by Wolter and Timlin (1998) from sea-level pressure, zonal and meridional components of the surface wind, sea surface temperature, surface air temperature and total cloudiness fraction of the sky. An average MEI for each hydrological year in the period 1951-2005 was obtained from published bimonthly MEI values (Wolter, 2007).

3.3.2.   Analyses

A series of non-parametric tests were carried out on several parameters that were derived from the temperature, rainfall, discharge and MEI dataset. For datasets which are not drawn from a population with specific statistical conditions, such as normal distribution, non-parametric tests are appropriate (McCuen, 2003). This is the case for the study data, which is not unusual for hydroclimatic data (Tilahun, 2006). To determine whether there was an overall trend in each time series studied, 2 trend analysis tests were used. The 1st trend analysis test, the Spearman rank test, is a simple measure of correlation between 2 data series and is a special case of the Pearson product-moment coefficient (Myers and Well, 2003). It has often been used for temporal analysis of climatic variables (e.g. Yu and Neil, 1993; González-Hidalgo et al., 2001). The Mann-Kendall test was also applied. This is a more elaborate test for identifying shift trends (e.g. Shao et al., 2009), and compares the relative magnitudes of the data, as opposed to their actual values (Gilbert, 1987). It has been widely used for assessing the significance of temporal trends in hydrological and climatic data (Hirsch et al., 1982), including in semi-arid regions (Zhang et al., 2008).

The time series from each parameter was also subjected to the Pettitt Test (Pettitt, 1979), which is often used to detect abrupt changes in hydrological series (e.g. Shao et al., 2009). The test determines the timing of a change in the distribution of a time series, known as a 'change point' (Zhang et al., 2008). The change point divides the series into 2 sub-series. The significance of the change point is then assessed by F- and t-tests on the change in the mean and the variance. The Pettitt test was used to identify change points in the time series, at a probability threshold of $p = 0.8$, followed by F- and t-tests at 2.5% significance level. This procedure has been used for identifying change points in hydroclimatic data in both humid and semi-arid environments (Tu et al., 2005; Ashagrie et al., 2006; Zhang et al., 2008). Tests were carried out using SPELL-stat v.1.5.1.0B (Guzman and Chu, 2004). To determine whether or not change points that were shown to be significant were also substantively important, the effect size was determined. The Cohen's d statistic was computed, with values of 0.8 and above considered to have a 'large' effect size (Cohen, 1988).

Using the Southern African hydrological year of October to September, the following time series were prepared from temperature records from Matopos Research Station (for which the longest time series in the region is available):
  - Annual average of daily maximum temperature in °C
  - Annual average of daily minimum temperature in °C
  - Annual maximum (extreme) of daily maximum temperature in °C
  - Annual minimum (extreme) of daily minimum temperature in °C

The following time series were prepared for rainfall for all climate stations:
  - Annual rainfall in mm a^{-1}
  - Number of wet days in d/a with rainfall >0.5 mm d^{-1}
  - Number of heavy rain days per year with rainfall >10 mm d^{-1}
  - Number of heavy rain days per year with rainfall >20 mm d^{-1}
  - Number of heavy rain days per year with rainfall >30 mm d^{-1}
  - Length of longest dry spell for the months November-March (days)
  - Number of 5 to 7 d dry spells for the months November to March (-)
  - Number of 8 to 14 d dry spells for the months November to March (-)
  - Number of more than 14 d dry spells for the months November to March (-)
  - Number of more than 20 d dry spells for the months November to March (-)
  - Total number of dry spells of 5 d or more for the months November to March (-)

Annual rainfall anomalies were prepared after the method of Hulme et al. (2001) for each station's annual rainfall series. The Thiessen polygon method was applied to estimate catchment rainfall; given the small number of rainfall stations per catchment, a more complex interpolation method was not justified. Catchment rainfall and discharge were compared for each discharge station at an annual time step and for selected stations at monthly and daily time steps.

The following time series were prepared for all discharge stations:
  - Annual runoff (mm a^{-1})
  - Annual maximum of monthly flows (m^3/month)
  - Days per annum no flow

- Maximum average daily flow recorded per year (m³/s)
- Maximum flood recorded (instantaneous flows) per year (m³/s)
- Runoff coefficient (-)

A simple risk analysis was carried out for each change point where a medium or large effect size had been determined. Risk here is defined as the probability of the system crossing an undesirable threshold in temperature, rainfall or discharge:

$$Risk = P(l > r) \tag{3.1}$$

Where P is the probability that load exceeds resistance, l is the load: the behaviour of the system under an external stress; r is resistance: the capacity of the system to overcome the load (Ganoulis, 2004)

For each parameter, a value was taken for resistance based on expert knowledge and related studies; each parameter represented a livelihood or an ecological risk. Thus for the selected parameters, the risk is the probability of occurrence of the below conditions:

- Temperature more than 3.1°C above the long-term mean for Bulawayo, shown to result in reduction of maize yield by 16% (Dimes et al., 2009)
- Rainfall under 450 mm a-1, the minimum rainfall recommended for rain-fed maize production (Shumba and Maposa, 1996; Hoffmann et al., 2002)
- Less than 60 and less than 30 wet days per year: representing 1 wet day per 3 days and per 6 days, respectively, in the 6-month rainy season
- Less than 20 days per year with rainfall of over 10 mm d^{-1}, often considered the baseline minimum of heavy rainfall events (Lebel et al., 2000; Douville et al., 2001). Heavy rainfall events are very important for runoff generation in an environment with high interception: a threshold of 5 mm d^{-1} has been suggested for the study area (Love *et al.*, 2010a).
- More than 10 days per year with rainfall of over 20 mm d$_{-1}$, which could result in waterlogging or flooding
- At least 1 dry spell per year lasting over 14 d: dry spells of this length are considered critical in reducing the yield of maize (Lal, 1997; Magombeyi and Taigbenu, 2008).
- At least 1 dry spell per year lasting over 20 d: more extreme than the preceding risk measure
- Annual runoff less than 22 mm a^{-1}: this is the mean annual runoff of the Mzingwane Catchment (Görgens and Boroto, 1997; MCC, 2009)
- Days of no flow more than 219 d/a: this shows the river to be flowing less than 40% of the time.
- Maximum average daily flow recorded per year under 5·m³/s and maximum instantaneous flood recorded per year under 20 m³/s: the occurrence of sufficiently large floods annually is important for hydro-ecological processes such as habitat maintenance (King and Louw, 1998). The values selected are indicative as the ecological significance of a specific flow volume or return period is specific to each river, and a detailed analysis is beyond the scope of this paper.

3.4. Results

3.4.1. Temperature

Figure 3.1 summarises the minimum and maximum daily temperature trends observed for 4 locations in the study area, with peaks corresponding with the occurrence of El Niño events (high positive MEI). Similar trends were observed at the 4 locations, but our focus will be on the Matopos Research Station location, which had the longest data set, some 20 years longer. The annual average of maximum daily temperatures at Matopos Research Station (Table 4) showed a significantly increasing trend (Spearman rank test). The MEI itself also shows a significantly increasing trend (Spearman rank test).

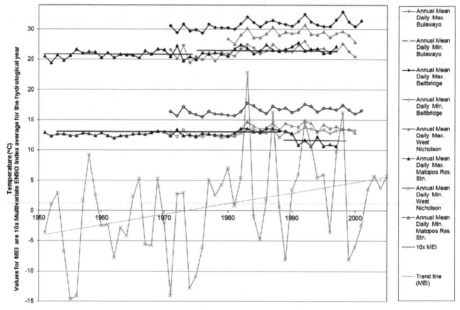

Figure 3.1. Annual means of daily temperature maxima and minima for 4 stations in the northern Limpopo Basin, Zimbabwe, contrasted with the Multivariate ENSO Index (MEI data: annual averages derived from Wolter, 2007), 1950/51 to 2000/01. The different sub-series for Matopos Research Station (horizontal lines show the sub-series means) are separated from each other by the change points identified in Table 3.3. For station locations see Figure 1.1.

Temperature parameters recorded at Matopos Research Station show change points in 1975 and the mid to late 1980s, with the average daily maximum temperature rising from 1975 and the annual average of maximum daily temperatures rising from 1986 (Table 3.3). A drop in minimum temperatures is recorded from around 1990 onwards.

Table 3.3. Results of Pettitt test on temperature parameters recorded at Matopos Research Station

Parameter	Significant change (t-test)	Date of Change Point	Effect size (Cohen's d)
Annual mean of the daily maximum temperature (°C)	Rise	September 1975	1.1
Annual maximum of the daily maximum temperature (°C)	Rise	September 1986	1.0
Annual mean of the daily minimum temperature (°C)	Drop	September 1990	2.9
Annual minimum of the daily minimum temperature (°C)	Drop	September 1989	2.6

3.4.2. Rainfall

The high inter-annual variability of rainfall in the northern Limpopo Basin is similar to that reported for much of Zimbabwe (Twomlow *et al.*, 2006). The influence of the ENSO phenomenon is shown by the correspondence of MEI peaks with unusually low rainfall for seasons such as 1982/83, 1986/87, 1991/92 and 1997/98 (Figure 3.2). For almost all stations there is a strong negative correlation between total annual rainfall and MEI: Mzingwane Dam (r=0.43), Bulawayo (r=0.46), Matopos Research Station (r=0.50), Filabusi (r=0.56), Mbalabala (r=0.56), Kezi (r=0.63), Beitbridge (r=0.70) and West Nicholson (r=0.71). Cyclicity of 17 to 20 years is apparent in the 5-year moving averages and to a lesser extent in the annual rainfall anomalies (Figure 3.3).

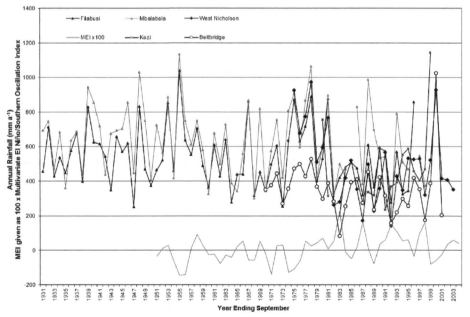

Figure 3.2. Annual rainfall of selected climate stations in the northern Limpopo Basin, contrasted with the Multivariate ENSO Index (MEI data: annual averages derived from Wolter, 2007), 1931/32 to 2003/04. For station locations see Figure 1.1.

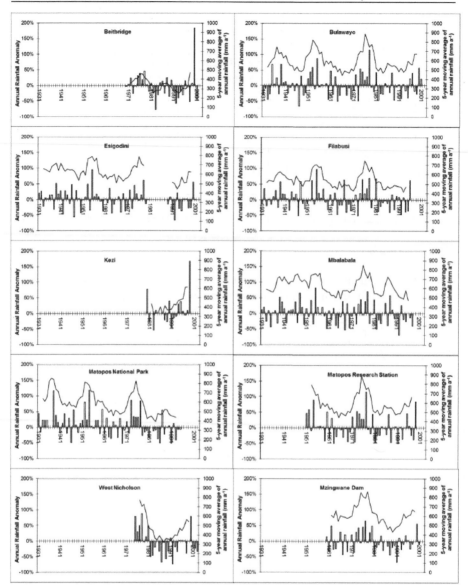

Figure 3.3. Annual rainfall anomalies (against 1961-1990 mean) and 5-year moving averages for selected climate stations in the northern Limpopo Basin, Zimbabwe, from 1930/31 to 2000/01.

Spearman rank and Mann-Kendall tests determined that only a limited number of parameters showed a significant drying trend for the whole time series (Table 3.4). These include the total annual rainfall and days of heavier rainfall at Matopos National Park, as well as the number of longer dry spells at Filabusi and Matopos Research Station. The Pettitt test identified numerous change points, with a significant difference in means confirmed by t-test. The majority showed a medium size effect ($0.3 < d < 0.8$) and several showed a large size effect ($d > 8$) (Table 3.4). Data for Bulawayo and Esigodini did not show significant change points and are not displayed. The majority of these showed a change point to a drier regime from around 1980, with a smaller number of change points that showed a change to a drier regime in the 1960s, as well as around 1980 (see Figure 3.4 for examples of the differences identified in Table 3.4). F-tests showed that most change points also showed a significant decrease in the variance of the rainfall parameter in question. All stations except Bulawayo and Matopos Research Station showed change to fewer wet days. Three stations showed change to an increased number of dry spells, a decreased annual rainfall or both. Only 1 station showed a change to fewer heavy storms. None of the stations showed any change to any wetter regime.

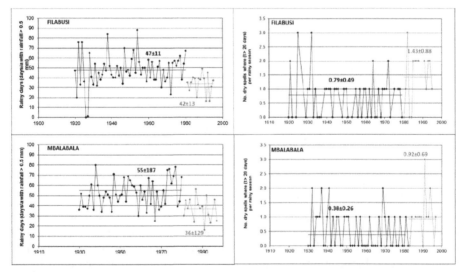

Figure 3.4. Examples of change points suggesting a possible change from a wetter to a drier rainfall regime: number of rainy days per year (left) and dry spells longer than 20 d (right) recorded at Filabusi (top) and Mbalabala (bottom). The different sub-series (as shown by horizontal lines) are separated from each other by the change points shown in Table 3.4, identified by Pettitt test and t-test. See Figure 1.1 for locations.

3.4.3. Discharge

The Pettitt test identified numerous change points, with a significant difference in means confirmed by t-test for discharge data obtained from 16 locations in the basin (Table 3.5). The majority of change points had a large size effect ($d > 0.8$)Most stations on tributaries of the Mzingwane, Thuli and Shashe Rivers show a change point to a drier regime from around 1980 (see Figure 3.5 for examples of the differences identified in Table 3.5). Most change points also showed a significant decrease in the variability of the parameter in question, confirmed by F-test –

although only about half of the stations show a decline in coefficient of variation. Almost all stations showed a change to a smaller number of days of flow and lower annual maximums of average monthly flows.

Many stations showed a consistently drying trend across the full time series for some parameters, including absolute declines in annual runoff at 6 stations, confirmed by Spearman rank and Mann-Kendall tests (Table 3.5). Data for B60 and B87 did not show significant change points and are not displayed. However, data from B87 showed continuous drying trends across the full time series, based on Spearman tests on the annual runoff, annual maximum of monthly flows, maximum daily flow per year and maximum flood recorded. Change points detected in B77 occur close to gaps in the time series and these results are also not displayed.

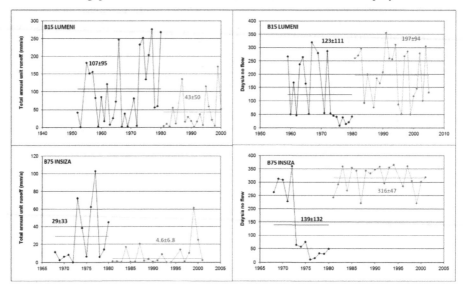

Figure 3.5. Examples of change points suggesting a possible change from a wetter to a drier discharge regime: total annual runoff (left) and number of days without flow per year (right) recorded at B15 Lumeni (top) and B75 Insiza (bottom). The different sub-series (as shown by black or grey lines) are separated from each other by the change points shown in Table 3.5, identified by Pettitt test and t-test. See Figure 1.1 for locations.

3.4.4. Rainfall-discharge relationships

Only a few discharge stations showed a clear relationship between annual runoff and annual catchment rainfall (Figure 3.6 and Table 3.6); these are predominantly tributaries of the Shashe and Thuli Rivers in the west: B77, B78, B80 and B83 (see Figure 1.1). All of these stations showed change to a drier regime in or around 1980 with at least 1 parameter showing an absolute declining trend across the time series (Table 3.5), although which parameter showed such a decline is not constant between the stations.

Change over time in the relationship between rainfall and runoff is shown by temporal change in the annual runoff coefficients (Table 3.6). Three stations show a significant decline in runoff coefficient, across the whole time series, 4 stations show a change to lower runoff coefficients in the early 1980s, and 1 station shows such a change in the mid 1980s. Examination of rainfall-runoff relationships at sub-annual time steps did not give acceptable levels of correlation, probably due to one of the following factors: differences in rainfall and discharge measurement days, delays between rainfall and discharge peaks at a short time step, or the high spatial variability of rainfall in Zimbabwe.

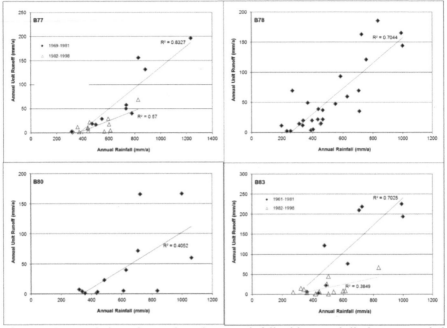

Figure 3.6. Correlation of annual catchment rainfall with annual discharge recorded at the 4 discharge stations in the northern Limpopo Basin, Zimbabwe that show a high correlation: those where $r > 0.6$. They are all from the west of the study area (see Figure 1.1 for locations). r = correlation coefficient.

3.4.5.   Risk analysis

Considering those parameters for which a change point with a medium or large effect size had been determined, the most important climatic risk observed is the increased probability of dry spells, with the probability of dry spells over 14 d and over 20 d rising to above 0.7 in many stations (Table 3.7). The probability of the number of rainy days per year falling below 60 and below 30 also increased substantially after each change point (mainly in 1979/80). There is no substantive risk of temperature exceeding the 3.1°C increase threshold and little increase in risk of annual rainfall levels dropping below 450 mm a^{-1} in the stations studied. The probability of heavy storms declined at the 2 stations for which statistically significant change was observed. There is a high risk of low annual runoff occurring. The probability of runoffa study catchment's annual runoff in mm a^{-1} being less than 22 mm a^{-1} (the overall average annual runoff on the Mzinganwe Catchment) doubled at most stations after the change points (Table 3.8). At all stations except B80, the probability of the river flowing less than 40% of time has increased substantially since the change points. In most cases, the probability of lower maximum flows and floods has increased.

Table 3.7. Risk analysis of climate parameters, comparing probability in series before and after identified change points (Table 3.4). For parameters where more than one change point was identified at a given station, the change point with the large effect size was selected.

Risk	Filabusi Series 1	Filabusi Series 2	Mbalabala Series 1	Mbalabala Series 2	Mzingwane Dam Series 1	Mzingwane Dam Series 2	Matopos Series 1	Matopos Series 2
Temperature 3.1°C above average							0.00	0.00
Rainfall under 450 mm a^{-1}					0.20	0.21	0.17	0.39
Less than 30 wet days per year	0.07	0.27	0.02	0.27	0.00	0.05	0.00	0.12
Less than 60 wet days per year	0.85	1.00	0.67	1.00	0.40	0.74	0.79	0.88
Less than 20 d/a with rainfall > 10 mm							0.54	0.06
More than 10 d/a with rainfall > 20 mm					0.35	0.16	0.33	0.06
At least 1 dry spell per year lasting more than 14 d	0.86	0.89					0.79	0.91
At least 2 dry spells per year lasting more than 14 d	0.50	0.89					0.55	0.73
At least 1 dry spell per year lasting more than 20 d	0.58	0.79	0.44	0.71				
At least 2 dry spells per year lasting more than 20 d	0.10	0.57	0.08	0.14				

Table 3.8. Risk analysis of discharge parameters, comparing probability in series before and after identified change points (Table 6). For parameters where more than one change point was identified at a given station, the change point with the large effect size was selected.

	Risk	Annual runoff < 22 mm a-1	Annual maximum of monthly flows < 1 m³/month	No flow > 219 d/a	Maximum average daily flow recorded per year < 5 m³/s	Maximum flood recorded (instantaneous) per year < 20 m³/s
B30 Mzingwane	Series 1			0.47		0.07
	Series 2			0.88		0.27
B11 Ncema	Series 1		0.12	0.29		0.36
	Series 2		0.17	0.68		0.45
B13 Nkankezi	Series 1			0.54		0.15
	Series 2			0.90		0.39
B61 Inyali	Series 1	0.38	0.69	0.73	0.50	
	Series 2	0.82	0.91	0.91	0.92	
B74 Jama	Series 1	0.18	0.55	0.70		0.36
	Series 2	0.68	0.82	0.92		0.73
B75 Insiza	Series 1	0.62	0.15	0.38		
	Series 2	0.90	0.65	1.00		
B15 Lumeni	Series 1	0.26	0.25	0.30	0.19	0.22
	Series 2	0.54	0.41	0.46	0.32	0.54
B78 Zgalangamate	Series 1	0.39	0.72	0.94	0.39	0.44
	Series 2	0.71	1.00	1.00	0.71	0.79
B80 Maleme	Series 1	0.17	0.17		0.00	0.00
	Series 2	0.92	0.67		0.82	1.00
B83 Mtshelele	Series 1	0.25	0.13	0.17	0.00	0.14
	Series 2	0.80	0.75	0.55	0.60	0.71

3.5.   Discussion

3.5.1.   Stationarity of the climatic and hydrological time series

The general trends in the time series, as identified by Spearman rank and Mann-Kendall tests, observed in the northern Limpopo Basin and the significant change points indicate:
-   The relationship between rainfall and temperature and ENSO (Figures 3.1 and 3.2); see also Alemaw and Chaoka (2006)
-   An increase in maximum daily temperatures (Table 2.2); see also New et al. (2006)
-   A decline in rainfall (Table 3.4 and Figure 3.4) with high inter-decadal variability (Figure 3.3); see also Trenberth et al. (2007)

Visual inspection suggests that a 3- to 5-year cyclicity may exist in rainfall (Figure 3.3), similar to and Jury and Mpeta (2005) and possibly related to El Niño. This cycle could be contained within a 17- to 20-year cyclicity in rainfall (Figure 3.3) which could be similar to Tyson (1986) and Mason et al. (1997). Thus, there is possibly a shorter cycle within a longer cycle, although this has not been tested statistically.

The vast majority of climate stations showed a change towards a drier climate from the early 1980s (Table 3.4) – the notable exception being Bulawayo, which is at a higher elevation and actually outside the Limpopo Basin. Decreased land rainfall from this period onwards was reported at global level by Trenberth et al. (2007). Decline in rainfall reflecting a new rainfall regime, as opposed to a gradual decrease, has also been reported from western Australia (Hope et al., 2006).

A corresponding change to a drier regime was also observed for most discharge stations (Table 3.5), notably from the early 1980s in the stations upstream of major dams (Mzingwane, Shashe and Thuli tributaries; see Figure 1.1 for locations). Such change in runoff regime has not been previously reported in south-eastern Africa. In 6 of the 16 stations (there is no consistent geographical pattern), there was a significant decline in annual runoff for the full time series. The change to a regime of relatively lower or less frequent rainfall and discharge demonstrates that water resource availability has shown a decline in the northern Limpopo Basin in the last half-century, even if trends in parameters such as annual rainfall are obscured in many places by the observed cyclicity.

3.5.2.   Causes of change in rainfall and discharge

The decline in rainfall and other drying trends in rainfall parameters (Table 3.4) could be related to:
-   The increasing frequency and severity of El Niño events (Alemaw and Chaoka, 2006).
-   Changes due to global warming: the more pronounced increase in number of dry spells with more limited changes in total rainfall is similar to those predicted from global circulation models by Christensen et al. (2007) for Southern Africa. The median precipitation response predicted is zero for December to May rainfall with a 13% decline in September to November rainfall (in southern Zimbabwe there is often very little rainfall during

September and October). The number of dry spells are predicted to increase in most subtropical areas.

- Possible effects of land use change on moisture recycling. Large areas of Zimbabwe, including parts of the study area, have experienced significant population growth, agricultural intensification and loss of forest cover in the last 50 years (e.g. Mapedza et al., 2003; Zinyama and Whitlow, 2004). However, changing sea surface temperatures (and thus ENSO) are thought to be more important than changing land use patterns in controlling warm season rainfall variability and trends in Southern Africa (Christensen et al., 2007). This needs further research.

- Observed cyclicity in rainfall: Figure 3.4 shows that the early 1980s are at the lower point of the 5-year moving average of annual rainfall. However, the cyclicity exhibited would suggest another change point (to a wetter regime) would be expected by the late 1990s. Such a change could not (yet) be observed, but may become apparent once more recent data become available.

The change to a drier regime in many discharge stations (Table 3.5) could be related to the following factors:

- Declining rainfall, as observed in Fig. 5, resulting in declining runoff generation. This explanation is favoured by the observed co-incidence of change points to decreasing rainfall and decreasing discharge in the early 1980s.

- Increased upstream withdrawals due to small dam construction. The Zimbabwe Government carried out an extensive dam-building programme; an estimated 15 000 small dams were built in Zimbabwe in the early to mid 1980s (e.g. Jewsbury and Imevbore, 1988; Chimomwa and Nugent, 1993).

- Post-1980 increases in agricultural water use, which will have occurred due to increased production since independence in 1980, at both smallholder and commercial scale (Thirtle et al., 1993), as well as due to changing population distribution, especially in areas which had experienced population displacement during the civil war (Zinyama and Whitlow, 2004).

- Possible effects of land-use change on hydrology: increasing population density and conversion of forest or rangeland to agricultural land, which has occurred in Zimbabwe over the past 50 years, see above, can cause a decrease in runoff (e.g. Calder et al., 1995) – although it can also cause an increase in runoff (e.g. Bari and Smettem, 2004).

3.5.3. Rainfall-runoff relationships

It should be noted that correlation of runoff with catchment rainfall will have been weakened by poor rainfall station distribution in the study area. Only a few discharge stations showed a good correlation (B77, B78, B80 and B83; see Table 7). These stations are all from the west of the study area (see Figure 1.1), where small dams are fewer, farm sizes are larger and crop production is minimal (Surveyor General of Zimbabwe and Forestry Commission, 1996); thus abstractions by farmers are likely to be lower in these areas. Seven discharge stations spread across the study area (B15, B40, B61, B74, B77, B83 and B87, Table 7) have declining runoff coefficients, showing a possible change in the runoff generation regime. Field studies, using experimental catchments with rainfall measured from several stations within the catchment, rather than interpolated from the national rainfall stations used in this study, might assist in further understanding of this problem.

3.5.4. *Risk analysis*

The principal risks posed by the changing hydroclimatic patterns observed are: risk of rainfed crop failure, due to an increasing probability of dry spells over 14 days and 20 days in length and an increasing probability of the number of rainy days falling below 30 days and below 60 days per year(Table 3.7), and risk to water supply (for agriculture and domestic use) from an increasing probability of annual runoff dropping below the Mzingwane Catchment average (Table 3.8). The observed changes in temperature and total annual rainfall at some stations do not appear to impose risk at this time. Changing flow regimes, with a higher probability of smaller-sized floods, may have ecological implications, but this cannot be assessed without a flood analysis.

3.6. Conclusions

There is a complex variety of factors that influences the temporal distribution of rainfall and discharge in southern Africa, some of which result in cyclicity. This complexity may be part of the reason that hydroclimatic studies using different statistical techniques make different findings. For example, Mazvimavi (2010) did not detect significant change in extreme high or low rainfall in 40 stations across Zimbabwe, using the Mann-Kendal test and quantile regression analysis. However, this study has shown that, in terms of the statistical tests selected in this study, total seasonal rainfall and annual discharges in the northern Limpopo Basin, Zimbabwe, have shown notable declines since 1980 at many measuring stations. The evidence from the time series analyses shows a declining rainfall trend that appears to be related to the increased incidence and severity of El Niño events and to changes associated with global warming. Discharge has been reduced by the declining rainfall input, but also by declining discharge from the headwater catchments, likely due to impoundments in small dams and increased abstractions for irrigated crop production reducing the volume of discharge from the runoff generated (especially in the north and east of the study area) . If rainfall continues to decline, it is likely that a multiplier effect will be felt on discharge, due to non-linear processes resulting in more than proportional declines in discharge (De Wit and Stankiewicz, 2006), as well as increased agricultural water abstractions by farmers in the headwater catchments to compensate with irrigation for the reduced rainfall.

It has been seen that since 1980 water resource availability in the study area has declined, both in terms of total annual water available for storage (i.e. declines in annual rainfall, annual runoff) and in terms of the frequency of water availability (i.e. declines in number of rainy days, increases in dry spells, increases in days without flow). Although construction of dams can mitigate inter-annual variability and the declining frequency of water availability, the decline in total annual water available for storage will decrease the yield of dams. These trends will further stress urban water supplies in the study area, notably those of Zimbabwe's second-largest city of Bulawayo, which already experiences chronic water shortages (Gumbo, 2004) and suggest that the construction on the Mzingwane River of the proposed Glassblock Dam (MCC, 2009), located between Mbalabala and West Nicholson (see Figure 1.1) should be preceded by a new feasibility study. Since the original analysis was carried out before 1980 (MCC, 2009), updating the yield analysis to consider the current trends is important. Taken together, these downward trends in water resources availability pose a challenge to that water supply development for south-western Zimbabwe (including any water exports from Zimbabwe). Comparative analysis of water resource availability trends in neighbouring basins should be considered in the light of possible alternatives to sourcing major water supplies from the Mzingwane Catchment and the Limpopo Basin, such as the long-awaited Matabeleland-Zambezi water carrier.

Drought and subsequent crop failure is becoming an all too common feature of south-eastern Africa (Richardson, 2007). This study has shown a rising risk of crop failure, due to an increasing probability of dry spells longer than 14 days occurring. Declining water resource availability in the region has had, and will undoubtedly continue to have, dramatic impacts on staple food production and food security (Du Toit and Prinsloo, 2001; Stige et al., 2006). The declining frequency of water availability will make the situation worse, and it has been shown that, at least for some parts of the study area, there is an increasing risk of annual runoff below the long-term catchment average. Unfortunately, many strategies to address hunger in southern Africa are water intensive, and some have been shown to result in water use conflicts (Love et al., 2006a). Such initiatives should be complemented by interventions that maximise the use of the existing scarce water resources. These include soil water conservation techniques (e.g. Mupangwa et al., 2006), rainwater harvesting (e.g. Mwenge Kahinda et al., 2007), improving the management of existing irrigation schemes (e.g. Samakande et al., 2004) and the use of alternative crops – such as switching from maize to sorghum – when a drier season is to be expected based on the condition of the ENSO phenomenon (Stige et al., 2006).

4. Rainfall-interception-evaporation-runoff relationships in a semi-arid catchment, northern Limpopo Basin, Zimbabwe[*]

4.1. Abstract

Characterising the response of a catchment to rainfall, in terms of the production of runoff vs. the interception, transpiration and evaporation of water, is the first important step in understanding water resource availability in a catchment. This is particularly important in small semi-arid catchments, where a few intense rainfall events may generate much of the season's runoff. The 30 km^2 ephemeral Zhulube Catchment in the northern Limpopo Basin was instrumented and modelled in order to elucidate the dominant hydrological processes. Discharge events were disconnected, with short recession curves, probably caused by the shallow soils in the Tshazi subcatchment, which dry out rapidly, and the presence of a dambo in the Gobalidanke subcatchment. Two different flow event types were observed, with the larger floods showing longer recessions being associated with higher (antecedent) precipitation. The differences could be related to: (i) intensity of rainfall or (ii) different soil conditions. Interception is an important process in the water balance of the catchment, accounting for an estimated 32 % of rainfall in the 2007-2008 season but as much as 56 % in the drier 2006-2007 season. An extended version of the HBV model was developed (designated HBVx), introducing an interception storage and with all routines run in semi-distributed mode. After extensive manual calibration, the HBVx simulation satisfactorily showed the disconnected nature of the flows. The generally low Nash-Sutcliffe coefficients can be explained by the model failing to simulate the two different observed flow types differently. The importance of incorporating interception into rainfall-runoff is demonstrated by the substantial improvement in objective function values obtained. This exceeds the gains made by changing from lumped to semi-distributed mode, supported by 1,000,000 Monte Carlo simulations. There was also an important improvement in the daily volume error (dV_d). The best simulation, supported by field observations in the Gobalidanke subcatchment, suggested that discharge was driven mainly by flow from saturation overland flow. Hortonian overland flow, as interpreted from field observations in the Tshazi Subcatchment, was not simulated so well. A limitation of the model is its inability to address temporal variability in soil characteristics and more complex runoff generation processes. The model suggests episodic groundwater recharge with annual recharge of 100 mm a^{-1}, which is similar to that reported by other studies in Zimbabwe.

---

[*] Based on:Love, D.; Uhlenbrook, S.; Corzo-Perez, G.; Twomlow, S.; van der Zaag, P. 2010a. Rainfall-interception-evaporation-runoff relationships in a semi-arid catchment, northern Limpopo Basin, Zimbabwe. *Hydrological Sciences Journal*, 55, 687-703. This paper received the Tison award from the International Association of Hydrological Sciences in 2012

4.2.   Introduction

Characterising the response of a catchment to rainfall, in terms of the production of runoff vs. the interception, transpiration and evaporation of water, is the first step in understanding water resource availability in a catchment. This is particularly important in semi-arid catchments, where a few intense rainfall events may generate much, or sometimes most, of the season's runoff (e.g. Lange and Leibundgut, 2003) and where spatial and temporal variability of rainfall can be high (e.g. Unganai and Mason, 2002). An understanding of the hydrological processes involved in a catchment is a basic requirement for integrated water resources management planning (e.g. Uhlenbrook et al., 2004). In southern Africa, where environmental and water stress is increasing (Nyabeze, 2004; Sivakumar et al., 2005), this type of understanding is essential in building resilience to large or catastrophic environmental changes and in developing trade-offs between food and economic production and ecosystem services (Falkenmark et al., 2007). It is also important for addressing broader humanitarian and development needs, through the many water intensive interventions that have been proposed by development agencies and projects (Love et al., 2006a).

For meso-catchments (scale of approximately $10^1 - 10^3$ km^2; Blöschl and Sivapalan, 1995) internal heterogeneities are very important (Didszun and Uhlenbrook, 2008). Evaporation processes, including interception, play a controlling role in runoff generation (Bullock, 1992) and interception is a major driver of the magnitude and speed of catchment response to rainfall, especially for semi-arid catchments with limited rainfall frequency and depth, and especially for smaller storm events (Seyam et al., 2000; Smith and Rethman, 2000; Beven, 2002; De Groen and Savenije, 2006; Tsiko et al., 2008). Despite this, many modelling studies either ignore interception, or consider it part of a lumped evaporation parameter (e.g. Smith & Rethman, 2000).

For semi-arid catchments, properties such as soil depth and permeability are also important, with percolation beginning much faster in catchments with shallower soil profiles (Chesson et al., 2004). Shallower soils dry more rapidly, which can reduce connectivity between discharge events (Farmer et al., 2003) – although in seasonal wetlands the narrow profile above an impermeable clay layer may remain saturated throughout the rainy season (McCartney et al., 1998). During high intensity rainfall, a soil's infiltration capacity can be rapidly reached, leading to overland flow as the initial phase of surface runoff. This process has been shown to become more dominant with increasing aridity in a number of study sites (Lange and Leibundgut, 2003) and increasing land degradation (Martinez-Mena et al., 1998). This emphasises the importance of differences in soil properties and land cover.

In data-poor regions, with ongoing declines in hydrological networks, hydrological prediction remains a challenge (Sivapalan et al., 2003). Where networks can be improved or extended, this is of course valuable – even a small number of discharge measurements can improve our understanding of a catchment or the performance of a model (Seibert and Beven, 2009). Such observations can be used to improve our understanding of the scientific basis of hydrology and of catchment response, which is a fundamental requirement for predictions in ungauged basins, one of the oldest and most critical tasks in hydrology – which has received increasing prominence under the International Association of Hydrological Scientists' PUB (Prediction in Ungauged Basins) initiative.

In this study, the response to rainfall of a meso-catchment in the semi-arid northern Limpopo Basin is studied and modelled. This is the first attempt at process-based hydrological modelling in the transboundary northern Limpopo Basin – although such studies have been carried out in headwater catchments of the Save and Zambezi Basins (e.g. McCartney *et al.*, 1998; Mugabe *et al.*, 2007). The objective of the study was to elucidate the dominant hydrological processes in the catchment. A dynamic, semi-distributed model was developed to analyse the rainfall-interception-evaporation-runoff relationships.

4.3.   Methods

4.3.1.   Study area

The selected study catchment, Zhulube, is a tributary of the upper Mzingwane River, located 87 km south-east of Bulawayo, Zimbabwe (Figure 4.1). Mean annual rainfall for the nearest climate station (Filabusi, 4 km from the northern edge of the catchment) was 555 mm a^{-1} for the period 1921-2006. The catchment is covered mainly by mixed *Hyparrehenia* grassland and *Bracyhstegia* woodland, with Highveld forest prominent in the northern Tshazi subcatchment (23 km^2) and degraded land prominent in the southern Gobalidanke subcatchment (7 km^2). Rainfed fields occupy the downstream portion of the catchment (Figure 4.1). The Tshazi subcatchment consists of steep, forested hills (altitude 1 000 to 1 200 m amsl; slope 3 to 20 %), with thin, rocky lixisols, whilst the Gobalidanke subcatchment is a shallow valley (altitude 1 000 m amsl; slope 1.3 %) between granite inselbergs (altitude 1 000 to 1 100 m amsl; slope 4 to 12 %), underlain by luvisols (Figure 4.2). The centre of the valley is a dambo, defined as a grass-covered, treeless seasonal wetland on hydromorphic soils (Wright, 1992), with a thin sandy horizon above a clay layer, superimposed on deeper sandy soil (Von der Hayden and New, 2003). Dambos are common in headwater streams on the Zimbabwean Highveld (McCartney *et al.*, 1998). The Tshazi subcatchment has eight tributaries with catchment areas over 1 km^2, whilst the Gobalidanke subcatchment has four. Tributaries in both subcatchments only flow during and immediately after some storm events.

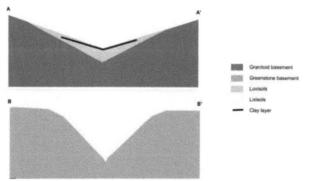

Figure 4.2. Schematic cross-sections through the Zhulube Catchment, based on geological mapping by Tunhuma *et al.* (2007) and field observations in gullies and artisanal mine workings. AA' is in the Gobalidanke Subcatchment and BB' in the Tshazi Subcatchment (see Figure 4.1). Not to scale.

4.3.2. Field Data

The catchment was instrumented with rainfall rain gauges as shown in Figure 4.1. These were read daily, and daily rainfall for each subcatchment was computed by use of Thiessen polygons, from the 2005-2006 rainy season onwards. All eight gauges were available for the 2007-2008 rainy season, but only five for the 2006-2007 season and four for 2005-2006. A composite gauge (v-notch and broad crest) was constructed at the catchment outlet in December 2006 and readings were taken daily at 0800 and 1600 hrs, with some additional readings taken manually during large storms. The latter procedure was instituted due to the repeated failures of auto-logging pressure transducers and the remote location of the gauge: the field assistant reported to the gauge during large storms and recorded the discharge level every ten minutes. Quality control of observations was made. A theoretical rating equation was used.

A daily antecedent precipitation index was determined for the time series, after the method of Casenave and Valentin (1992):

$$APL_i = (APL_{i-1} + P_{i-1})e^{-\alpha t} \qquad (4.1)$$

Where API_i is the antecedent precipitation index for day i (mm d^{-1}), P_i is the rainfall recorded for day i (mm d^{-1}), α is a weight, assigned the value of 0.5 as widely used in semi-arid regions, and t is the time which has elapsed since the last rainfall event prior to day i.

From this a daily antecedent effective precipitation index was determined by replacing rainfall in equation (4.1) with rainfall less interception (estimated according to equation (4.7) below).

4.3.3. Multiple Regression Rainfall-Runoff Model

A spreadsheet-based multiple regression model was prepared to study the rainfall-runoff relationships in the Zhulube catchment. The model generated effective rainfall and simulated runoff through multiple linear regression for different interception thresholds and by considering different number of days of antecedent rainfall. In addition to other observations, the model was used to calibrate the interception threshold D, which determines the maximum amount of water that can be stored on the land and vegetation surface (De Groen and Savenije, 2006; Seyam et al., 2000).

4.3.4. Estimation of Interception and Evaporation

Daily reference evaporation was calculated using the Hargreaves formula, equation (4.2) (Allen et al., 1998). It has been suggested that this formula, which is based on temperature and radiation, provides the best input for streamflow simulations in semi-arid areas (Oudin et al., 2005).

$$E_0 = 0.0023(T_{mean} + 17.8)(T_{max} - T_{min})^{0.5}(0.408Ra) \qquad (4.2)$$

Where E_0 is reference evapotranspiration (mm d^{-1}) incorporating interception evaporation, transpiration and soil evaporation, T_{mean} is mean daily temperature (°C), T_{max} is maximum daily temperature (°C), T_{min} is minimum daily temperature (°C) and Ra is daily insolation (J m^{-2} d^{-1} x10^6). Temperature and radiation data were taken from the West Nicholson climate station, which was the closest station with such data (48 km from the study site).

Potential evapotranspiration for different land covers was derived using equation (4.3):

$$E_i = Kc_iE_0 \qquad\qquad (4.3)$$

Where E_i is the potential evapotranspiration for land cover i (mm d^{-1}) and Kc_i is the crop coefficient for land cover i (-). The crop coefficient selected for maize varies according to the stage of development; values from Allen *et al.* (1998) and FAO (2013b) for East Africa were used. For other land covers, South African equivalents were selected, varying monthly (Table 4.1).

Table 4.1. Crop coefficients used for different land cover types, varying by season. For distribution of the land cover types, see Figure 4.1.

Land Cover, this study	Woodland: highveld	Mixed grassland and woodland	Mixed grassland and woodland (degraded)
South African equivalent	Woodland (indigenous tree/bush savanna) [a]	Mixed bushveld [b]	Veld in poor condition [c]
Jan	1.14	1	0.79
Feb	1.14	1	0.79
Mar	1.14	0.93	0.79
Apr	1.14	0.86	0.64
May	1	0.71	0.29
Jun	1	0.64	0.29
Jul	1	0.57	0.29
Aug	1	0.64	0.29
Sep	1.07	0.79	0.43
Oct	1.14	0.93	0.57
Nov	1.14	0.93	0.71
Dec	1.14	1	0.79

Source: [a] Jewitt (1992); [b] Schulze and Hols (1993); [c] Schulze *et al.* (1995). The original crop coefficients were derived for use with pan evaporation data (Schulze *et al.*, 1995). These were converted for use with reference evapotranspiration data by dividing the original crop coefficient with the pan coefficient (taken as 0.7).

From these data and the mapped land cover distribution (Figure 4.1) potential evapotranspiration at subcatchment level was calculated:

$$E_c = X_1 E_1 + X_2 E_2 + \cdots + X_n E_n \tag{4.4}$$

Where E_c is the potential evapotranspiration at subcatchment level (mm d^{-1}) and X_i is the area fraction of the subcatchment under land cover i (-).

Since interception is a threshold process (Seyam et al., 2000; Savenije, 2004; Fenicia et al., 2008), daily interception, can be calculated using the equation below

$$I_i = minimum(P_i, D) \tag{4.5}$$

Where I_i is the interception for day i (mm d^{-1}), P_i is the rainfall recorded for day i (mm d^{-1}) and D is the interception threshold (mm d^{-1}) (Savenije, 2004). However, if some rainfall was intercepted on the previous day, and the amount of rainfall intercepted was more than could be evaporated on that day, some moisture will remain in interception storage until the next day, thus decreasing the available volume of interception storage for that day. It is assumed that moisture in interception storage at subcatchments level evaporates with reference to potential evaporation (E_c). A daily interception time series was thus determined for each subcatchment:

$$S_i = maximum(I_i - Ec_i, 0) \tag{4.6}$$

$$I_i = minimum(P_i, D - S_{i-1}) \tag{4.7}$$

Where S_i is the stored interception at the end of day i (mm) and E_{ci} is the potential evaporation at subcatchment level on day i (mm d^{-1}). Equation (4.7) could also be extended to consider interception storage carried from more than one day antecedent.

Daily interception values (from equation 7) were then used to derive daily transpiration and soil evaporation, which were not further separated:

$$E_{sc} + T_c = maximum(E_c - I_i, 0) \tag{4.8}$$

Where E_{sc} is soil evaporation (mm d^{-1}) and T_c is net transpiration (mm d^{-1}), both for the weighted land covers as combined in equation (4.4), at subcatchment level as indicated by the subscript c.

4.3.5.   The HBVx Model and Model Application

The HBV (Hydrologiska Byråns Vattenbalansmodell) family of models, whilst developed and applied initially in Sweden, have also been used in semi-arid and arid countries such as Australia and Iran (Masih et al., 2008; Oudin et al., 2005). The application of HBV in Zimbabwe has previously been limited to the humid subtropical climate of eastern and northern Zimbabwe (Líden and Harlin, 2000).

"HBV light" (Seibert, 2002) consists of four routines: snow (not used in this study), soil moisture, response and routing. The model can be run lumped, or semi-distributed. In the latter mode, it is only the soil moisture which can be parameterised in a distributed manner (considering up to three vegetation zones and a selected number of elevation zones) and none of the input data (precipitation, temperature, evaporation) is distributed. Two improvements of the model structure have been made with the development of HBVx. First, an interception volume is introduced, as per the previous section. Second, all of the routines can be run in parameterised and semi-distributed mode, through the designation of two or more separate sub-basins. The basic equations for the linear reservoirs are given in Figure 4.3.

The soil routine is based on two parameters: *FC* (mm), which defines the maximum soil moisture storage or field capacity; this can be emptied by evaporation. β (-) defines the non-linear function that computes the amount of infiltration water that goes into the runoff generation routine (toRGR) and the amount that stays in the soil routine to fill up the soil moisture storage (SM):

$$\frac{toRGR}{P} = \left(\frac{SM}{FC}\right)^{\beta} \tag{4.9}$$

Parameter *LP* (-) is the soil moisture ratio threshold below which the actual subcatchments evaporation does not reach E_c, due to moisture stress (Seibert, 2002; Uhlenbrook et al., 2004). This reduces the soil evaporation and transpiration value (i.e. nett of interception per equation (4.8) at sub-catchment level) when SM/FC is less than LP.

Only the 2007-2008 season data was used, since there were insufficient data points in the 2005-2006 season. Initial calibration of the lumped HBV model without interception was carried out to explore the parameter space. The parameter ranges are shown in Table 4.2, and were selected based on field observations and experience in application of HBV to other catchments.

Table 4.2. Parameter ranges for all runs performed in the calibration of HBVx. MAXBAS was set at 1 and LP at 0.7 throughout all runs. See Figure 4.3 for the role of each parameter in the model. No value is given for *D* as the runs were performed either without interception storage (*D* = 0 mm) or with the single interception storage value derived from multiple regression.

Parameter	FC (mm)	β (-)	UZL (mm)	K_0 (d^{-1})	K_1 (d^{-1})	K_2 (d^{-1})	PERC (mm d^{-1})
Minimum	10	1	10	0.25	0.1	0.0001	2
Maximum	150	5	100	1.00	0.7	0.0050	5

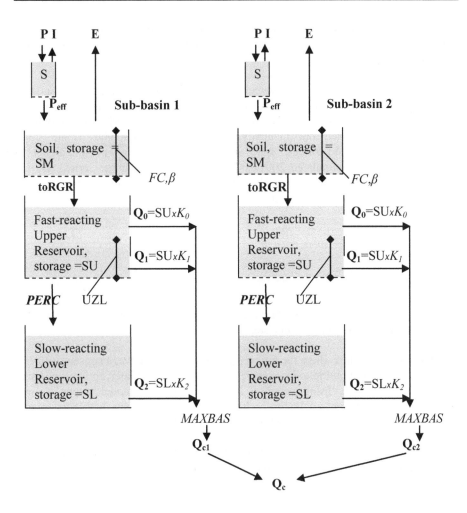

Figure 4.3. Schematic diagram of the HBVx model structure in semi-distributed mode with interception routine.

Parameters: P = Precipitation, S = Interception storage (capacity = D), E = Soil evaporation and transpiration - see equation (4.8), I = Interception, P_{eff} = Effective rainfall, toRGR = Moisture transferred to runoff generation routine, UZL = Threshold for start of overland flow, Perc = Percolation, Q_0 = Overland flow, Q_1 = Discharge from saturated soil or shallow groundwater, Q_2 = Discharge from deep groundwater, Q_c = Total discharge from catchment, MAXBAS = Routing parameter. All parameters mm d^{-1}, except UZL and FC (mm) and MAXBAS (-). Fluxes are shown in bold and model parameters in italics. Note that, with the exception of Q_c, all fluxes and parameters are different for each of the two subcatchments.

For the semi-distributed model setups, the linear storage coefficients K_0, K_1 and K_2 were higher and the soil storage parameters *UZL* and *FC* generally lower for the Gobalidanke subcatchment. This was done in order to represent the effects of the higher proportion of degraded land in the Gobalidanke subcatchment (see Figure 4.1), from which faster overland and near-surface flows were observed. This field observation is also supported by the typical soil profile in the Gobalidanke subcatchment: a near-surface clay layer, expected to result in soil storage drying more rapidly.

A stepwise model concept improvement approach (Fenicia *et al.*, 2008) was then used to evaluate the separate and incremental benefits of incorporating the interception routine and of the semi-distributed parameterisation. Extensive calibration was carried out manually, supported by 1,000,000 Monte Carlo simulations, within the parameter space selected (Table 4.2), to (i) obtain a good fit of the shape of the simulated 2007-2008 rainy season discharge series, and (ii) maximise the objective functions. The selected objective functions were the Nash-Sutcliffe Coefficient (C_{NS}) and mean volume error (dV_d) (mm a^{-1})

$$C_{NS} = 1 - \frac{\sum_{i=1}^{n}(Q_{obs,i}-Q_{sim,i})^2}{\sum_{i=1}^{n}(Q_{obs,i}-\overline{Q_{obs}})^2} \qquad (4.10)$$

$$dV_d = \frac{365 \times \sum_{i=1}^{n}(Q_{obs,i}-Q_{sim,i})}{n} \qquad (4.11)$$

Where Q_{obs} (mm d^{-1}) is the observed discharge, Q_{sim} (mm d^{-1}) the simulated discharge and n the number of time steps i (days) in the simulation.

For those model setups where the interception routine was active, a constant value of D obtained from the multiple linear regression was used, and interception flux I_i was calculated in HBVx, using equation (4.5). For lumped model setups, I_i was calculated at catchment scale for each timestep, but for semi-distributed model setups it was calculated independently at subcatchment scale.

A simple sensitivity analysis was carried out for the best parameterisation for the eight parameters that were varied during the calibration. The 10 % elasticity index (e_{10}) was used as per Cullmann and Wriedt (2008):

$$e_{10} = \frac{Out_1-Out_0}{0.1 \times Out_0} \qquad (4.12)$$

Where Out_0 is the initial model output being studied and Out_1 is the model output after the parameter in question has been increased or decreased by 10 % (both were done).

4.4.   Results

4.4.1.   Field Data

Rainfall observations showed high spatial and temporal variability, with annual totals for the 2005-2006 and 2007-2008 seasons close to the long term average annual rainfall at Filabusi, but total rainfall for the 2006-2007 season was well below this average (Table 4.3). The latter observation is probably related to a moderate positive ENSO anomaly which was recorded during that season (Logan *et al.*, 2008).

Table 4.3. Annual rainfall statistics from field observations in Zhulube Catchment, contrasted with the long term average for Filabusi. For locations see Figure 4.1.

Season	Total annual rainfall (mm a^{-1})	Number of rainy days (-)
2005-2006	528	49
2006-2007	289	31
2007-2008	592	57
Filabusi average, 1921-1996	555	51

Examination of the daily rainfall and discharge data from Zhulube for the 2006-2007 season (from February 2007 onwards, when the gauge was operational) does not show a consistent pattern in either the initiation of discharge nor in the source (between the two sub-catchments). The discharge events are highly disconnected, with no observable recession curve (Figure 4.4a). A further factor contributing to the disconnected nature of the discharge events is the tendency for soils observed elsewhere to become hydrophobic during droughts, thus decreasing infiltration (Beven, 2002).

During the 2007-2008 season, it can be seen that when there is a difference between the subcatchments rainfall values, rainfall in the Gobalidanke Subcatchment gives an apparently stronger discharge response than rainfall in the Tshazi Subcatchment: compare for example the discharge recorded on 16.02.2008, in response to 19 mm rainfall in Gobalidanke Subcatchment with the lack of discharge recorded on 01.03.2008 to 20 mm rainfall in Tshazi Subcatchment (Figure 4.4b) – despite a higher API of 1.2 mm d^{-1} on the second date to 0.2 mm d^{-1} on the first date.

Two peak types can be seen (Table 4.4): (i) floods lasting less than one week and (ii) flows with recession exceeding two to three weeks. The latter group of floods are associated with much higher antecedent precipitation ($Q/API \geq 1$), although the runoff coefficients are variable. If there is any difference in the time of initiation of response, it is not observed and therefore is likely to be at a sub-daily scale (this could also not be observed in the bi-daily raw data). The very sharp recession curves could be caused by (presumed) low antecedent soil moisture and lack of baseflow to this ephemeral river system. This is exacerbated in the Gobalidanke Subcatchment by the shallow soil horizon (above the impermeable clay). Discharge was found to follow the pattern of antecedent precipitation (Figure 4.5).

Table 4.4. Characteristics of flood events recorded in the Zhulube Catchment, 2007-2008 season.

Date of maximum flow	Total discharge recorded Q (mm)	Total rainfall recorded P (mm)	Duration of event (days)	Event runoff coefficient Q/P (-)	$Q/API*$ (-)
04.12.2007	0.33	19.43	2	0.02	0.07
16.12.2007	6.23	131.77	6	0.05	0.16
18.12.2007	10.91	47.69	6	0.23	0.06
29.12.2007	3.66	44.90	19	0.08	0.91
10.01.2008	5.96	60.93	13	0.10	5.47
25.01.2008	9.52	105.71	17	0.09	2.3E+04

* Antecedent precipitation index

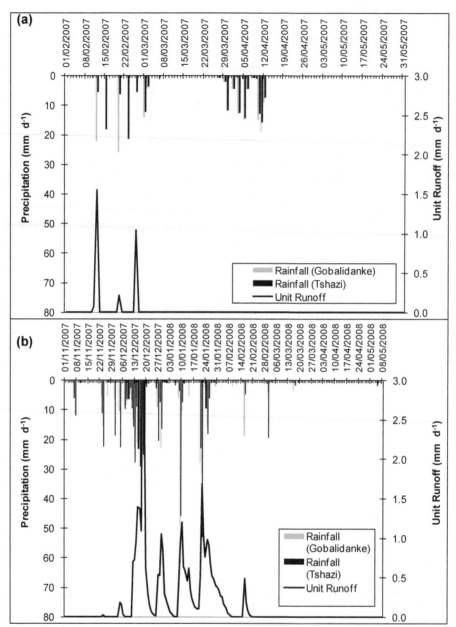

Figure 4.4. Observed rainfall, disaggregated by subcatchments, and discharge, Zhulube Catchment: (a) 2006-2007 rainy season; (b) 2007-2008 rainy season.

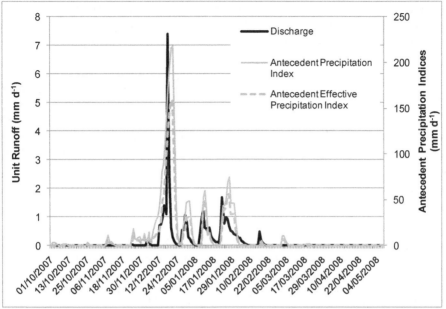

Figure 4.5. Discharge compared with the antecedent precipitation indices, as per equation (1).

4.4.2. Multiple Regression Rainfall-Runoff Model

The results of the model suggest that the consideration of only the previous day's rainfall (as opposed to longer periods of antecedent rainfall) influence the observed discharge (Table 4.5). Values of between 2 mm d^{-1} and 6 mm d^{-1} were used for the interception threshold (D); this being the range proposed by De Groen and Savenije (2006). Sensitivity to interception threshold is low, though a threshold of 5 mm d^{-1} gives marginally better results.

Runoff was simulated using an interception threshold of 5 mm d^{-1} (as the threshold giving marginally better correlation; Table 4.5) and marginally better results were obtained using catchment rainfall, as compared to subcatchments rainfall (Table 4.6). The general runoff dynamics were simulated well, except the largest flood of 18.12.2007. Discharge was over-simulated during the start and end of the rainy season (Figure 4.6), probably due to the model not considering (soil) storage.

Table 4.5. Correlation coefficients from multiple linear regression of discharge and effective rainfall, Zhulube Catchment. D = interception threshold, mm d^{-1}.

Antecedent rainfall considered in model (days)	Correlation coefficient (-)				
	D = 2	D = 3	D = 4	D = 5	D = 6
0	0.5797	0.5758	0.5709	0.5652	0.5581
1	0.7087	0.7087	0.7097	0.7098	0.7067
2	0.7452	0.7478	0.7497	0.7507	0.7509
3	0.7599	0.7631	0.7656	0.7674	0.7685
4	0.7600	0.7632	0.7657	0.7675	0.7687
5	0.7778	0.7813	0.7843	0.7868	0.7888
6	0.7788	0.7824	0.7854	0.7881	0.7902
7	0.7788	0.7825	0.7855	0.7881	0.7902

Table 4.6. Comparison of observed discharge in the Zhulube Catchment with simulated discharge using different rainfall series and an interception threshold of 5 mm d^{-1}. Q_{simZ} = discharge simulated using Zhulube Catchment rainfall

	Observed Discharge Q_o	Simulated Discharge Q_{simZ}
Mean (mm d^{-1})	0.21	0.23
Standard deviation (mm d^{-1})	0.65	0.48
Correlation Coefficient R (-)		0.77
Nash-Sutcliffe Coefficient C_{NS} (-)		0.59

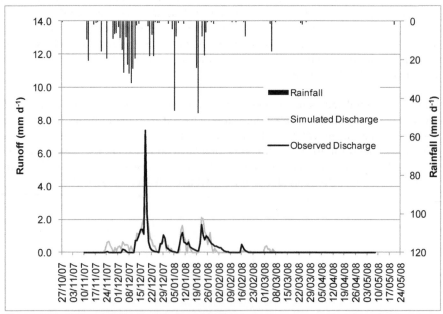

Figure 4.6. Observed discharge, Zhulube Catchment, compared to that simulated from catchment rainfall using the multiple regression rainfall-runoff model with an interception threshold of 5 mm d^{-1}. Note the over-simulation of discharge during the earliest and latest parts of the season.

4.4.3. Interception Estimations

Given that the results of the multiple regression model favoured a memory of only one day, equation (4.6) was used as stated above. An interception threshold value of 5 mm d^{-1} was used as before. In the Zhulube Catchment, it can be seen that much of the early and late rainfall was intercepted in the 2007-8 season. Interception values decrease slightly on days following heavy rains (Figure 4.7). Interception accounted for 56 % of rainfall in the 2006-2007 season, and 32 % of the rainfall in the 2007-2008 season. Transpiration and soil evaporation values for the Gobalidanke Subcatchment are (often) lower than for the Tshazi Subcatchment (Figure 4.8), due to the minimal forest cover and extensive degraded areas in the former.

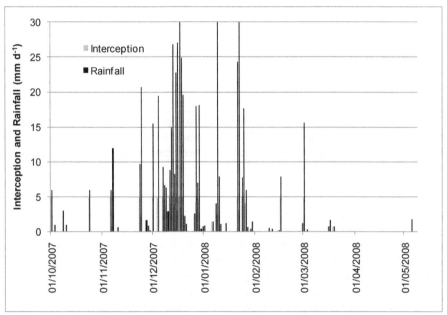

Figure 4.7. Daily rainfall and interception, Zhulube Catchment, 2007-2008 season. Total season rainfall was 592 mm and total season interception was 167 mm. The effect of equation (6) can be seen as interception values decrease slightly on some days following consecutive heavy rains.

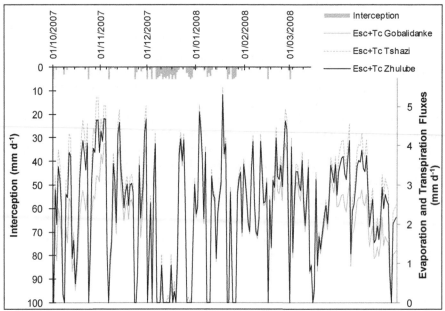

Figure 4.8. Computed daily interception and transpiration plus soil evaporation (E_{sc}+T_c), Zhulube Catchment, 2007-8. Soil evaporation and transpiration values for the Gobalidanke Subcatchment are generally lower than for the Tshazi Subcatchment.

4.4.4. HBVx Modelling

The automatic calibration of the lumped model setup without interception storage achieved only relatively low Nash-Sutcliffe coefficients (C_{NS}), and the mean daily volume error (dV_d) values were always below zero (Figure 4.9), indicating an over-simulation of the observed discharge. Changing from lumped to semi-distributed mode did not produce any real improvement in the objective functions, with dV_d remaining negative. However, introduction of interception storage did: C_{NS} of 0.503 for the model in semi-distributed mode with interception storage, compared to C_{NS} of 0.290 without (Table 4.7). The value of dV_d improved from -44.9 mm d^{-1} without interception storage to 6.9 mm d^{-1} with. The semi-distributed set up with interception storage performed slightly better than the lumped setup with interception storage. The best parameterization of each model set-up is shown in Table 4.7 and Figure 4.9. Varying the interception threshold D from that determined in the multiple regression consistently produced poorer Nash-Sutcliffe coefficients and mean daily volume errors.

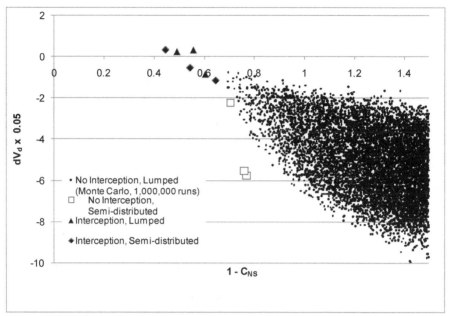

Figure 4.9. Dot-plot of performance of different model set-ups and parameterisations of the HBVx models. The Nash-Sutcliffe coefficient is displayed as $(1-C_{NS})$ so that both objective functions would be zero for a perfect simulation. The mean daily difference is expressed as $(dV_d \times 0.05)$. For parameterisation, see Table 4.6; for objective functions see equations (10) and (11).

Comparing the two best parameterisations within the semi-distributed model setup with interception, SI1 and SI2, both have high FC and UZL storage parameters (soil storage and upper groundwater zone storage respectively) and simulate discharge mainly through Q1 (which should approximate flow from shallow groundwater), with no Q_0 (which should approximate overland) flow. SI1, with slightly lower K_1 value (meaning slower flow Q_1) gives the best performance in terms of objective functions, although the fit is not so good (see Figure 4.10a).

The model is not able within one parameterisation to produce a good fit to both the short, intense discharge peaks (e.g. 18.12.2007) and the slower peaks with developed recessions (e.g. 10.01.2008, 25.01.2008).

It can be seen that the thresholds FC and UZL are not reached in either subcatchments. Discharge recedes as the upper zone dries out and nearly ceases once the upper zone is dry (compare Figures 4.10a and 4.10b, e.g. on 24.12.2007). For the floods which peaked on 10.01.2008 and 25.01.2008, recession over several days is not simulated by the model. The model simulates an extremely small Q_2 flow (which approximates baseflow from deep groundwater); according to field observations this should be zero for most of the year.

Figure 3.7 ...

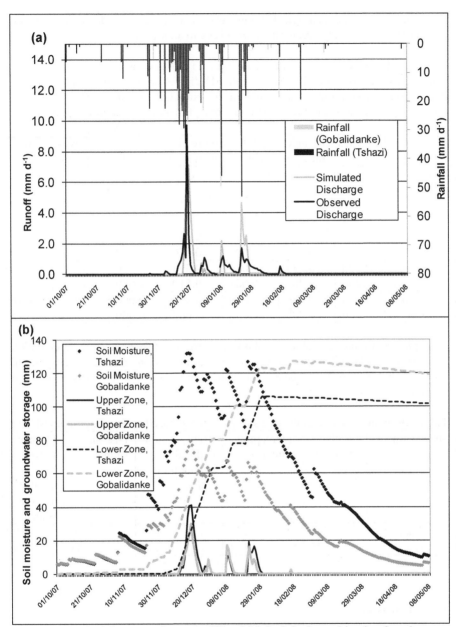

Figure 4.10. Model outputs for setup SI1, which had the best objective function values. (a) Observed and simulated discharge and rainfall. (b) Simulated soil moisture and groundwater storage for the two subcatchments.

The sensitivity analysis (Table 4.8) showed that the total simulated discharge, Q_1 flow (flow from shallow groundwater, the largest portion of simulated flow) and recharge of the lower groundwater zone are particularly sensitive to *PERC* (maximum daily flow to lower groundwater zone) and *FC*, with mean soil moisture sensitive mainly to *FC*. After PERC, the model is most sensitive to β, and sensitivity to the hydrological storage coefficients are lower. The model output appears more sensitive to soil and geological parameters. Sensitivity to a decrease in model parameter was generally greater than sensitivity to an increase.

Table 4.8. Local sensitivities of model outputs to model parameters, setup SI1, calculated using equation (12). Since setup SI1 did not produce any overland flow (Q_0), no sensitivity could be calculated for this output. Parameters UZL and FC are both storage threshold parameters and only one of them can be a limiting factor at a time.

Model output	Elasticity index (e_{10}), first figure given is for 10 % decrease in the model parameter, second figure is for 10 % increase						
	K_0	K_1	K_2	PERC	UZL	FC	β
Total Q_{sim}	0.000	0.401	0.119	0.818	0.000	1.802	0.456
	0.000	0.423	0.118	0.606	0.000	1.490	0.292
Total Q_1	0.000	0.258	0.000	0.663	0.000	0.595	0.125
	0.000	0.294	0.000	0.570	0.000	0.574	0.098
Recharge	0.000	0.233	0.082	0.594	0.000	0.101	0.231
	0.000	0.266	0.081	0.510	0.000	0.122	0.195
Mean soil moisture	0.000	0.000	0.000	0.000	0.000	1.175	0.169
	0.000	0.000	0.000	0.000	0.000	1.193	0.143

A simple water balance (Table 4.9) shows the prominent role played by interception. The Tshazi subcatchment shows greater transpiration and soil evaporation (presumably due to the greater woodland coverage) and less discharge than Gobalidanke subcatchment.

Table 4.9. Water balance of the two subcatchments, based upon model setup SI1.

Subcatchment	Rainfall (mm a^{-1})	Interception (mm a^{-1})	Transpiration and soil evaporation (mm a^{-1})	Discharge (mm a^{-1})	Change in soil and aquifer storage (mm a^{-1})
Gobalidanke	506	173	167	70	120
Tshazi	575	164	254	53	206

4.5. Discussion

4.5.1. Observed Rainfall and Discharge Characteristics

Observed discharge events were disconnected at catchment level, with short to very short recession curves. This is exacerbated by the high spatial variability in rainfall. In the Tshazi Subcatchment, the disconnected flows are probably caused by the shallow soils which dry out rapidly, resulting in little baseflow and reduced connectivity between events (Farmer *et al.*, 2003). In the Gobalidanke Subcatchment, the presence of an impermeable clay layer limits percolation and thus baseflow. In common with other studies in Zimbabwe (e.g. Bullock, 1992; Bullock and McCartney, 1995; McCartney, 2000) and Zambia (Von der Hayen and New, 2003), the presence of a dambo in Gobalidanke is likely to contribute to discontinuous discharge during the rainy season, as water is retained and transpired by the wetland. There is also no extension of discharge into the dry season, again as seen in studies elsewhere in Zimbabwe. Even after a good season, discharge drops to zero very quickly after a flood event.

Flow events were either short, intense peaks lasting less than one week or flows with slower recession, lasting two to three weeks. Variation in flow processes across a rainy season has been shown in many sites in Sub-Saharan Africa (Dubreuil, 1985). The difference in the type of flood in the study site can be associated with differences in antecedent precipitation (Table 4.4) and could be related to two factors: (i) Rainfall intensity (which was not measured), with more intense storms producing the flash floods. (ii) Change in soil conditions over the course of the rainy season. For example soil crusting, the formation of thin dense near-surface layer of low hydraulic conductivity, typically occurs before or at the start of the rainy season (Hopmans et al., 2007) or during vegetative droughts (Beven, 2002); the 2006-2007 season was a drought. Crusting would decrease infiltration and could lead to flash floods, especially at the beginning of the rainy season. Such behaviour is consistent with Hortonian (infiltration excess) stormflow. Floods later in the season, such as those in late January which were generated mainly in the Gobalidanke subcatchment, are consistent with saturation overland flow from a dambo (McCartney et al., 1998). Variation in soil parameters such as infiltration rate across a rainy season has been shown at two sites in the Zhulube Catchment by Ngwenya (2006): An ungrazed study site showed a change from an initial sorptivity (capillarity) control on infiltration rate to hydraulic conductivity control (Table 4.10).

Quality control of rainfall data showed that adjacent stations gave close results, and no major changes were noted between measurements by different observers. The composite gauge performed well at lower flows (using the V notch), but there was lower reproducibility of observations from the broad crest.

Table 4.10. Field measurements of infiltration rate taken from two sites in the study catchment during the 2005-2006 rainy season, using a tension infiltrometer. Data from study of Ngwenya (2006).

Date	Treatment	R^2 (cumulative infiltration : time)	R^2 (cumulative infiltration : time$^{0.5}$)	Conclusion
13.12.2005	Clay fenced	0.999	0.977	Sorptivity
	Clay grazed	0.991	0.997	Hydraulic conductivity
13.01.2006	Clay fenced	0.986	0.997	Hydraulic conductivity
	Clay grazed	0.993	0.994	Hydraulic conductivity
23.01.2006	Clay fenced	0.978	0.998	Hydraulic conductivity
	Clay grazed	0.994	0.993	Sorptivity?
28.01.2006	Clay fenced	0.989	0.996	Hydraulic conductivity
	Clay grazed	0.995	0.994	Sorptivity?
05.02.2006	Clay fenced	0.983	0.997	Hydraulic conductivity
	Clay grazed	0.999	0.981	Sorptivity?
10.02.2006	Clay fenced	0.977	0.999	Hydraulic conductivity
	Clay grazed	0.995	0.990	Sorptivity?
15.02.2006	Clay fenced	0.984	0.997	Hydraulic conductivity
	Clay grazed	0.986	0.996	Hydraulic conductivity
27.02.2006	Clay fenced	0.997	0.985	Sorptivity
	Clay grazed	0.993	0.941	Sorptivity
27.03.2006	Clay fenced	0.999	0.981	Sorptivity
	Clay grazed	0.988	0.993	Sorptivity
13.01.2006	Clay-loam fenced	0.988	0.999	Hydraulic conductivity
	Clay-loam grazed	0.983	0.994	Hydraulic conductivity
23.01.2006	Clay-loam fenced	0.983	0.985	Hydraulic conductivity
	Clay-loam grazed	0.997	0.989	Sorptivity
28.01.2006	Clay-loam fenced	0.978	1.00	Hydraulic conductivity
	Clay-loam grazed	0.891	0.992	Hydraulic conductivity
05.02.2006	Clay-loam fenced	0.997	0.992	Sorptivity?
	Clay-loam grazed	0.359	0.358	Sorptivity?
10.02.2006		Not measured		
15.02.2006	Clay-loam fenced	0.984	0.998	Hydraulic conductivity
	Clay-loam grazed	0.999	0.986	Sorptivity?
27.02.2006	Clay-loam fenced	0.985	0.997	Hydraulic conductivity
	Clay-loam grazed	0.998	0.982	Sorptivity?
27.03.2006	Clay-loam fenced	0.993	0.992	Hydraulic conductivity
	Clay-loam grazed	0.998	0.972	Sorptivity?

The two years of measurements from Zhulube already provide some insight into the response of this catchment. However, the variation in flood event type during the 2007-2008 season, and the great difference between that season and 2006-2007 show that constraining predictive uncertainties may be more complex in highly variable semi-arid catchments than the simpler case of perennial humid catchments (Seibert and Beven, 2009). Understanding of catchment dynamics would have been improved by gauging the two subcatchments. However, there were no suitable sites upstream of the confluence of the Tshazi and Gobalidanke streams. For future research, the findings of this study would be improved if supported by detailed measurement of soil moisture and groundwater levels. The experimental design included recording rainfall and discharge at sub-hourly timesteps. However, such results are not available due to equipment failure and theft.

4.5.2. Modelling results

The HBVx simulation satisfactorily showed the ephemeral, disconnected nature of the flows. The importance of incorporating interception into rainfall-runoff modelling is demonstrated by the substantial improvement in objective function values obtained when this was done in the model – exceeding the gains made by changing from lumped to semi-distributed mode (Figure 4.8). The relatively low Nash-Sutcliffe coefficients can be explained by the model failing to simulate the two different observed flood types differently. Shortcomings of the model include the inability to address temporal variability in soil characteristics and the exclusion of some storages which may sustain flow, such as bank and wetland storage. However, the experimental data is insufficient to support a more complex model set-up, with additional storages and fluxes.

The best HBVx simulation (SI1) suggests discharge driven mainly by flow from the fast-reacting upper reservoir (SU), similar to saturation excess overland flow, as expected in the Gobalidanke subcatchment but not the Tshazi subcatchment. It simulates the subsurface flow in a physically realistic manner, although there are no field data to compare the simulation with. The model suggests episodic groundwater recharge – see the increase in lower zone groundwater storage in Figure 4.9b – as reported by Butterworth et al. (1999). The estimated groundwater recharge of 100 mm a^{-1} (Figure 4.10b) is within the range reported from elsewhere in Zimbabwe by Larsen et al. (2002), but somewhat higher than those reported by Farquharson and Bullock (1992), Sibanda et al. (2009) or Wright (1992). However, it should be remembered that the modelled season (2007-2008) had higher than average rainfall (see Table 4.3).

4.6. Conclusions

The Zhulube Catchment has shown strong spatial variability in rainfall, and ephemeral, disconnected discharge events, even in a season with above-average rainfall. Two different flood types were observed, probably caused by different runoff generation processes, influenced by catchment antecedent soil moisture.

The extended HBVx model can satisfactorily model the ephemeral flow, and the minimal baseflow from deep groundwater, but does not appear to perform as well when catchment parameters or processes vary during the duration of a calibration interval. The best HBVx simulation that is supported by field observations in the Gobalidanke subcatchment suggested that discharge was driven mainly by flow similar to saturation excess overland flow. Hortonian overland flow, as interpreted from field observations in the Tshazi Subcatchment, was not simulated well.

Interception is an important process in the water balance of this semi-arid catchment, with the model suggesting that interception could account for 32 % of rainfall in the 2007-2008 season but as much as 56 % in the drier 2006-2007 season. The importance of interception is reflected in significantly improved performance of HBVx once this routine is introduced. However, it should be understood that the above finding is based upon the rainfall-runoff modelling carried out and not on actual field measurements of interception. There is a possibility of equifinality between interception and soil evaporation in the model.

For the future, understanding of Zhulube, and ephemeral, semi-arid catchments of this type, could be improved by a longer time series, and observations at higher temporal resolution. This would allow for a more complex model set-up and the evaluation of the variation of soil parameters in space and time.

5. Regionalising a meso-catchment scale conceptual model for river basin management in the semi-arid environment[*]

5.1. Abstract

Meso-scale catchments are often of great interest for water resources development and for development interventions aimed at uplifting rural livelihoods. However, in Sub-Saharan Africa IWRM planning in such catchments, and the basins they form part of, are often ungauged or constrained by poor data availability. Regionalisation of a hydrological model presents opportunities for prediction in ungauged basins and catchments. This study regionalises HBVx, derived from the conceptual hydrological model HBV, in the semi-arid Mzingwane Catchment, Limpopo Basin, Zimbabwe. Fifteen meso-catchments were studied, including three that were instrumented during the study. Discriminant analysis showed that the characteristics of catchments in the arid agro-ecological Region V were significantly different from those in semi-arid Region IV. Analysis of flow duration curves statistically separated catchments with flow more than two thirds of the time from those with flow less than 60 % of the time. Regionalised parameter sets for HBVx were derived from means of parameters from the three catchment groups and all catchments. The parameter sets that performed best in the regionalisation are characterised by slow infiltration with moderate/fast "overland flow". These processes appear more extreme in more degraded catchments. This suggests that benefits can be derived from conservation techniques that increase infiltration rate and from runoff farming. Faster, and possibly greater, sub-surface contribution to streamflow is expected from catchments underlain by granitic rocks. Calibration and regionalisation were more successful at the dekad (10 days) time step than when using daily or monthly data, and for the Group C catchments (flow over two-thirds of the time). However, none of the regionalised parameter sets yielded $C_{NS} \geq 0.3$ for half of the catchments. The HBVx model thus does offer some assistance to river basin planning in semi-arid basins, particularly for predicting flows in ungauged catchments at longer time steps, such as for water allocation purposes. However, the model is unreliable for more ephemeral and drier catchments. Without more reliable and longer rainfall and runoff data, regionalisation in semi-arid ephemeral catchments will remain highly challenging.

5.2. Introduction

Small to meso-scale catchments are often of great interest for water resources development (e.g. Mazvimavi, 2003; Niadas, 2005; Nyabeze, 2005), for environmental planning (Walker *et al.*, 2006) and for development interventions aimed at uplifting rural livelihoods (Ncube et al., 2010). In semi-arid areas of Sub-Saharan Africa, rising water demand and the challenge of frequent droughts creates a desire for water resources development and a requirement for integrated water

[*] Based on: Love, D.; Uhlenbrook, S.; van der Zaag, P. 2011. Regionalising a meso-catchment scale conceptual model for river basin management in the semi-arid environment. *Physics and Chemistry of the Earth*, 36, 747-760.

resources management planning in order to balance food security, other economic needs and the needs of the environment in the allocation and development of surface water flows (Peugeot et al., 2003; Love et al., 2006a). These challenges will grow worse with rising populations in most river basins, and the anticipated impacts of global warming leading to increased water scarcity (Fung et al., 2011) and making the need for IWRM planning more pressing.

However, many river basins suffer from limited data availability and more limited process knowledge (Bormann and Diekkrüger, 2003; Ndomba et al., 2008). In developing countries, many basins are ungauged (Mazvimavi et al., 2005). This lack of data constrains planning and can be a stumbling block to conflict resolution among users competing for scarce water resources (Nyabeze, 2000). Prediction of discharge and other hydrological characteristics of ungauged basins is therefore an important priority for water resources management – as well as for hydrological science – and was adopted by the International Association of Hydrological Sciences in 2002 as the Prediction of Ungauged Basins (PUB) research agenda (Sivapalan et al. 2003).

One approach to address these challenges is regionalisation, which provides methods to upscale small-scale (or meso-scale: scale of approximately $10^1 - 10^3$ km^2; Blöschl and Sivapalan, 1995) measurements to a large scale (or basin or regional scale: scale of $> 10^4$ km^2) model, or to outscale measurements from gauged catchments to ungauged catchments. For the purposes of hydrological modelling, key parameters to regionalise are those that represent processes, such as flow response decay time, and contribution of faster and slower flows to total discharge (Littlewood et al., 2002; Troch et al., 2002). These parameters are essential for process-oriented simulation, which is important in order to better understand the possible effects of different environmental influences (Ott and Uhlenbrook, 2004; Johst et al., 2008). Additionally, there is tension between adequate model parameterisation, in order to address heterogeneities within and between catchments, and rising model complexity and uncertainty (Beven, 1993; Lin and Radcliffe, 2006; Marcé et al., 2008) – and the challenges that limited data availability pose to the latter issues (Bormann and Diekkrüger, 2003).

Model parameter sets are calibrated through a series of models runs on catchments which have a long time series of data (Heuvelmans et al., 2004). Calibration can be against a single critical parameter (e.g. Nyabeze, 2005) or multi-criteria calibration of a parameter set (e.g. Seibert, 2000; Uhlenbrook and Leibundgut, 2002). Alternatively, a multivariate statistical approach can be used to determine the relationships between biophysical catchment descriptors and hydrological catchment parameters (e.g. Chiang et al., 2002; Mwakalila et al., 2002). No single regionalisation procedure has been developed that yields universally acceptable results (Ramachandra Rao and Srinivas, 2003) and all regionalisation methods require reliable and long-term data series. This can be problematic in semi-arid regions of Africa (Nyabeze, 2002; Mazvimavi, 2003), such as the Mzingwane Catchment, the portion of the northern Limpopo Basin that lies within Zimbabwe, in which 11 of the 30 secondary catchments are completely ungauged, there is high inter-annual and intra-annual variability in runoff and many ephemeral tributary catchments (Nyabeze, 2002; 2005; Love et al., 2010b).

For river basins with many ungauged catchments, or with limited data availability, regionalisation represents one of the possible approaches to addressing the lack of data required for planning purposes. An alternative approach is to estimate model parameters from catchment characteristics (Koren *et al.*, 2004; Kapangaziwiri and Hughes, 2008), although this also requires sufficient available data.

The Mzingwane Catchment Council, the stakeholder-based statutory authority for water resources planning in the catchment, needs to balance different sectoral water requirements and issue new water permits as demand changes and new areas are developed. For these purposes, an understanding of annual runoff is needed (Mzingwane Catchment Council and Zimbabwe National Water Authority, 2009).

Given the data constraints of the Mzingwane Catchment, it is helpful to be able to regionalise model parameters from better-understood catchments to poorly gauged and ungauged tributaries.

This study explores the challenges of regionalisation of widely-used box models in semi-arid catchments. The main objective is to regionalise one or more model parameter sets for the Mzingwane Catchment using HBVx, a model developed from HBV (Bergström, 1992; Seibert, 2002) in a gauged field study catchment (Love et al., 2010a). The regionalisation exercise will use field study data and historic data from those gauged tributary catchments within the national hydrological data network for which there is reasonable data availability. The second objective is to determine whether or not distinct groups of tributary catchments can be identified and separate parameter sets regionalised for each group. The third objective is to improve the understanding of catchment behaviour and relate it to catchment characteristics.

5.3. Methods

5.3.1. Study area

Seventeen meso-catchments within the northern Limpopo Basin were selected on the basis of the following criteria: (i) availability of discharge data. (ii) Development: the selected catchments were upstream of all major dams. (iii) Proximity to rainfall station(s): less than 50 km distance. (iv) Catchment scale: area 1,500 km ^{2}. (v) Catchment shape: the long axes of the selected catchments were less than 50 km in length, measured from catchment outlet to the most distant point on the watershed (Engeland *et al.*, 2006), in order to exclude long, narrow catchments liable to be highly diverse. The selected catchments are shown in Figure 1.1 and their data sources in Tables 5.1 and 5.2. Characteristics of the catchments are set out in Table 3. Fifteen catchments were used for calibration and regionalisation of model parameters and the other two for blind regionalisation: evaluating the performance of the regionalised parameter sets against catchments whose data had not been previously used in calibration.

Table 5.2. Instrumentation installed in field study site catchments

Catchment	Discharge stations	Climate stations
M27 Mnyabezi 27	Dam and limnigraph	7 rain gauges Class A evaporation pan
MSH Mushawe	Bridge and limnigraph	17 rain gauges Class A evaporation pan
UBN Upper Bengu	Dam and limnigraph	8 rain gauges Class A evaporation pan
Zhulube	Composite gauge (V-notch and broad crest)	14 rain gauges Class A evaporation pan

5.3.2. Data quality control

The quality of input data is of high importance since this influences both model performance and the parameter sets to be regionalised. There is a minimum quantity of input data required for model parameterisation and where input data is limited by missing values, problems can be created as it has been shown that where measurements are only available for some days, results may differ significantly depending upon which days measurements are available for (Seibert and Beven, 2009). The study areas have high spatial and temporal variability in rainfall, and runoff (Love *et al.*, 2010a; 2010b) which is likely to exacerbate this problem.

The time series were visually inspected, along with supporting materials such as the station files. The following exclusions were made for each station, in order to remove unreliable data: (i) Where rainfall or discharge data was missing for two months or more, the year was excluded. (ii) Where rainfall or discharge data was missing for two weeks or more during the months of November to April (rainy season), the year was excluded. (iii) Where a note has been made in the station file that readings were unreliable (e.g. due to siltation, security), the year was excluded. (iv) For dekad time series (10 days), dekads containing one or more days with missing rainfall or discharge data were excluded. (v) For monthly time series, months containing six or more days with missing rainfall or discharge data were excluded.

5.3.3. Catchment classification

The catchments were classified into groups, for separate calibration of the HBVx model parameters.

The first classification method was based upon a provisional classification, made using the standard Zimbabwean agro-ecological zones (Vincent and Thomas, 1960). The agro-ecological zonation takes into account climatic factors and soil types. This provisional classification was then refined by discriminant analysis. This is a multivariate analysis procedure for testing the statistical significance of a pre-existing (user-determined, not statistically-derived) classification. The canonical discriminant functions that separate classes from each other are derived, using a linear combination of the variables and the Mahalanobis distance (Chiang; 2002, Gordon et al., 2004). The technique which has proven effective in the classification of river basins, using catchment characteristics as the variables and proposed groups of catchments as the classes (Wiltshire, 1986; Chiang; 2002). For this study, eight catchment characteristics were selected to describe the main properties of the catchments, see Table 5.3 excluding the designation of agro-ecological region, which is the subject of the classification.

The second classification method used was comparison on monthly flow duration curves of the catchments using the Kolmogorov-Smirnov test. This is a non-parametric test used to fit the cumulative distribution function of a sample to a distribution function. (Loucks and van Beek, 2005). A two sample Kolmogorov-Smirnov test determines whether or not two sample groups come from statistically different populations (Panik, 2005). It has been used extensively in hydrology (Tate and Freeman, 2000; Niadas, 2005), including for the comparison of flow duration curves (Kileshye-Onema et al., 2006).

5.3.4. *HBVx model calibration*

For an explanation of the HBVx model adapted in this study, please see section 4.3.5.

The time series for the selected catchments were each calibrated at three time steps: daily, dekad (10 days) and monthly. Prior to each run, the model was initialised in order to better represent initial conditions (Noto et al., 2008). Initialisation was for one year (daily time step), 1.5 years (dekad) or two years (monthly). Calibration was carried out using 20 000 runs of the genetic algorithm method (Seibert, 2000). The genetic algorithm calibrations were repeated several times to confirm that similar parameter sets were derived from each calibration. The selected objective functions were the Nash-Sutcliffe Coefficient (C_{NS} – see equation 4.10) and mean volume error (dV_d – see equation 4.11)

5.3.5. *Regionalisation*

The parameters sets which were developed during the calibration processes were used to develop regionalised parameter sets for HBVx: the means for each parameter of the values generated from calibration of the group A catchments, group B catchments and group C catchments, at daily, dekad and monthly time steps.

The three best-performing regionalised parameter sets were then applied blindly to two catchments (B56 and B64) which had not been used in the classification and calibration exercises. This exercise could not be extended to more catchments, as the remaining catchments had been eliminated as having insufficient data quality.

5.3.6. *Uncertainty of input data*

The uncertainty in the model that may arise due to variability of rainfall data has been discussed. The sensitivity of catchment rainfall, the major input to the model, to spatial variability of rainfall within a catchment was evaluated using the 10 % elasticity index (e_{10} – see equation 4.12).

5.4.   Results

5.4.1.   Observed flows

The high inter-annual variability in runoff can be seen in Figure 5.1, especially in the more arid catchments B26, B78 and B90.A comparison of catchment area with runoff coefficient shows a clear negative relationship (Figure 5.2).

5.4.2.   Catchment classification

Discriminant analysis of the catchment classification by agro-ecological region shows a strong statistical basis for separating the catchments into those in Region IV and Region V (Table 5.6). The variables that form the strongest basis for the classification are the standard deviation of the annual rainfall and the land cover.

Table 5.6. Results of discriminant analysis of catchment characteristics.

Main results	
Groups	2
Variables	8
Cases	15
Group 1 (Region IV)	B11, B15, B30, B39, B60, B61, B74, B80, B83
Group 2 (Region V)	B26, B78, B90, M27, MSH, UBN
Wilk's Lambda	0.03267 (0 = perfect discrimination)
Tolerance of variables	
Variable	*Tolerance (0 = completely redundant)*
Geology	0.151
Soil	0.223
Land cover	0.356
Degradation	0.253
Rainfall mean	0.264
Rainfall standard deviation	0.452
Tenure	0.283
Slope	0.281

Two sets of flow duration curves were prepared for the selected catchments: monthly flow normalised against catchment area (unit flow, mm month^{-1}) (Figure 5.3a) and monthly flow normalised against mean monthly flow (-) (Figure 5.3b).

Visual inspection of the flow duration curves (Figure 5.3) suggests three groups (Table 5.7). The three Group A catchments are all in the drier south of the study area (Figure 1.1) – and correspond to Group 2 of the discriminant analysis (Table 5.5), that is, the catchments in Region V. The three Group C catchments are in the central northern area, but there are also Group B catchments in that area.

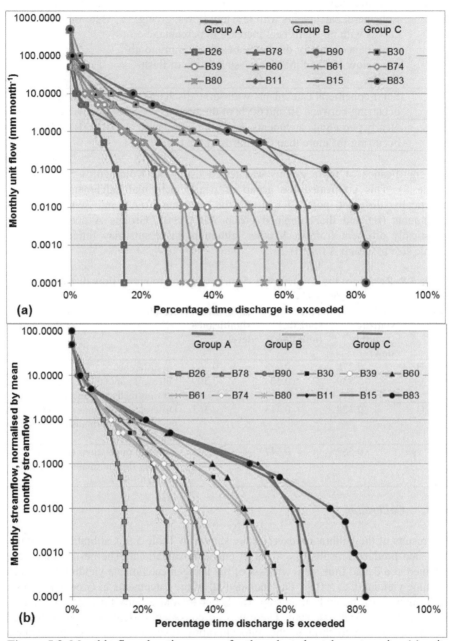

Figure 5.3. Monthly flow duration curves for the selected catchments, using (a) unit flow (monthly discharge values normalised by catchment area) and (b) monthly discharge values normalised to the mean for that catchment

Table 5.7. Catchment classification based on monthly flow duration curves.

Group	Characteristic flow duration curve	Catchments
A	Ephemeral catchments with few low flows, giving curves with no significant increase in percentage exceedance time for discharge below 0.1 mm month⁻¹ or below 10 % of the mean – suggestive of flash floods.	B26, B78, B90
B	Semi-ephemeral catchments with some discharge occurring between 30 and 60 % of the time	B30, B39, B60, B61, B74, B80
C	Nearly perennial catchments, with some discharge occurring for more than two-thirds of the time	B11, B15, B83

The significance of these groups was tested using the Kolmogorov-Smirnov test (Table 8). This confirmed the group of nearly perennial catchments (C) as a statistically different population from the ephemeral (A) and semi-ephemeral catchments (B) and the combined group AB (A+B). Groups A and B are not statistically different from each other, although this result may reflect the small sample size in Group A ($n = 3$).

Table 5.8. Results of the Kolmogorov-Smirnov test on monthly flow distribution, using C_α from Stephens (1974).

Case	Test statistic (monthly flow)	Test statistic (unit flow)	K_α at 0.10 significance level	Result
A vs. B	0.278	0.278	0.369	No difference
A vs. C	0.559	0.504	0.509	Different populations (for monthly flow normalised to mean only)
B vs. C	0.428	0.443	0.353	Different populations (for both monthly flow normalised to mean and unit flow)
AB vs. C	0.562	0.547	0.326	Different populations (for both monthly flow normalised to mean and unit flow)

5.4.3. Calibration

The results of the calibration exercise are shown in Table 5.9. Calibration at a daily time step produced results $C_{NS} > 0.3$ in only one catchment. The best results were obtained at a dekad time step, with six of the thirteen catchments yielding $C_{NS} > 0.4$ and nine yielding $C_{NS} > 0.3$. This included all of the catchments in Groups B and C. Performance at a monthly time step was better than at a daily time step but not as good as at the dekad time step. There was no consistent difference in the values of parameters generated through the autocalibration (available in the supplementary material) between the different groups, except for generally low FC values for Groups B and C.

Model performance was compared with several catchment characteristics. Higher C_{NS} values were associated with the more perennial catchments (R=0.40, negative correlation of performance to days of now flow), but the best correlation was with the proportion of degraded land in a catchment (R=0.59, negative correlation). Other catchment descriptors compared to model performance did not show correlation. The proportion of degraded land was found to be positively correlated to the fast runoff coefficient, often interpreted as overland flow K_0 (from the parameter sets developed during calibration; see supplementary material) at a daily time step (R=0.45) but not correlated with the intermediate runoff coefficient, K_1, and the slow runoff coefficient, K_2 (R=-0.18, R=-0.23, respectively). These are often interpreted as discharge from shallow groundwater and deep groundwater, respectively. The proportion of the catchment underlain by granitoid was found to be positively correlated to the coefficients for flow from the two sub-soil reservoirs K_1 and K_2 (from the parameter sets developed during calibration; see supplementary material) at a daily time step (R=0.49 for K_1; R=0.43 for K_2) but not to the coefficient for "overland" flow from the soil box (R=0.21 for K_0).

5.4.4. Regionalisation

The parameter sets used in regionalisation are shown in Table 5.10. The performance of each of these sets is shown in Table 5.11. The three field study catchments gave very poor results throughout, probably due to the short time series (1 or 2 years). The parameter set which performed best was "*dekadBC*", with 43 % of the series giving $C_{NS} > 0.3$. The second best was "*monthlyBC*", with 36 % of the series giving $C_{NS} > 0.3$. The third set was "*monthlyABC*", with 29 % of the series giving $C_{NS} > 0.3$. All of these parameter sets have low values for β, PERC 1 and moderate values for K_1 and FC. This is indicative of relatively slow infiltration and percolation with moderate to fast "overland flow". Blind regionalisation gave mixed results (Table 5.12), with each parameter set giving $C_{NS} > 0.3$ for only one of the two catchments at a given time step. Performance is better for B56, the less arid catchment. The sensitivity of catchment rainfall to spatial variability is clearly shown in Table 5.13: a 10 % change in rainfall of one rainfall station has a substantial effect on catchment rainfall. On a daily time step, the effect can be extreme if rainfall is only reported from one station. This can also be seen in Figure 5.4: the gauge at *E Nyathi* recorded heavy rainfall on 8 January, but the other gauges in the catchment recorded heavy rainfall the following day. Had only three or less gauges been used to represent rainfall in the catchment (as is the case for the catchments not instrumented in this study), the catchment rainfall could easily have been incorrectly estimated on either day.

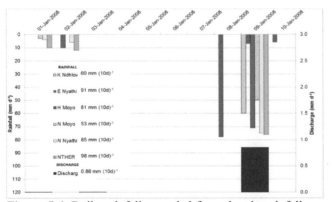

Figure 5.4. Daily rainfall recorded from the six rainfall gauges in the Mnyabezi 27 research catchment (M27) and daily discharge recorded from the catchment outlet for the first dekad (ten day period) of 2008. Total fluxes for the dekad are given in the key.

5.5. Discussion

5.5.1. Catchment classification

Comparison of the results of the discriminant analysis of catchment characteristics (Table 5.6) and the Kolmogorov-Smirnov test on the flow duration curves (Table 5.8) suggests that the agro-ecological region classification (Vincent and Thomas, 1960) can be related to catchment flow characteristics. The lack of statistical support for the separation between Groups A and B (Table 5.8) could be due to the small sample size of Group A or could suggest a limitation to agro-ecological region classification as a predictor of catchment behaviour. The distinction between Group C catchments and Groups A and B catchments (Table 5.7), based upon the flow duration curves, was statistically supported (Table 5.8).

5.5.2. Model performance

During both calibration (Table 5.9) and regionalisation (Table 5.11), model performance was poor at a daily time step, with the exception of the catchments instrumented during this study. The poor performance at a daily time step can be explained by two factors: First, the measurement day for runoff is 00:00 to 23:59 of the same day while the measurement day for rainfall is 08:00 to 07:59 of the next day. However, most rainfall in Zimbabwe occurs in thunderstorms (Mazvimavi, 2003), which generally occur in the afternoon, which should minimise the impact of this error.

Second, rainfall in Zimbabwe shows high spatial variability (Mugabe et al., 2007; Unganai and Mason, 2002), which means that the available climate stations may under-represent or over-represent rainfall which occurs in a given catchment, especially the larger ones, and especially at a daily time step (Mazvimavi, 2003). For many of the catchments, the only available rainfall data comes from gauges outside, but adjacent to, the catchment. Spatial variability is not as great at a dekad time step, as can be seen from the totals in Figure 5.4. Whilst a monthly time step will average out the spatial heterogeneity in rainfall data, it has the disadvantage that size and shape of discharge peaks are lost as discharge events lasting one or two days are averaged across the month. This variability is associated with uncertainty in catchment rainfall from a coarse network (Table 5.13).

Neither of these two factors applied to the catchments instrumented during this study as the measurement day was the same for rainfall and for runoff (08:00 to 07:59 of the next day) and the instrumented catchments had between 7 and 17 rainfall stations.

The Group C catchments, with flows occurring over two-thirds of the time are easier to simulate with the selected box model than the Group A and B catchments, with flow less than 60 % of the time. This was also the case for the blind regionalisation, where the less arid and less ephemeral catchment was simulated better. This could be related to the fact that flow in ephemeral catchments is highly unequally distributed in space and time (Lange, 2005). Ephemeral catchments have more threshold processes and many more discrete flow events, with large, short-term variations in discharge, which are more difficult to simulate (Johst et al., 2008).

Furthermore, the information content for a given length of time series is more limited for ephemeral catchments (Woolridge et al., 2003).

The arid zone environment thus imposes constraints to the utility of the HBVx model. The more continuous the discharge, i.e. fewer events and lower variation of the Group C catchments, the more suitable for simulation with HBVx . A large part of the challenge is being able to make satisfactory quantification of catchment rainfall (Hughes, 1995). This is likely similar with other conceptual box models.

5.5.3. *Implications for semi-arid zone hydrology*

The strong negative correlation between model performance and proportion of the catchment which is degraded suggests a strong influence of land degradation on flow processes. The fact that the proportion of degraded land is positively correlated to the fast runoff "overland flow" coefficient K_0 but not to the other flow coefficients suggests that land degradation can be linked to more rapid flow, especially overland flow. This is likely to largely be through the effect of loss of vegetation facilitating a fast response. Similar findings were made, for instance, by Lange and Leibundgut (2003) in the Sahel, where land degradation was associated with decreased infiltration and increased overland flow. Rapid and more episodic flow events tend to be more discrete and thus more difficult to simulate with a box model.

The correlation between the fraction of a catchment underlain by granite and the "groundwater" flow coefficients K_1 and K_2 compares well to the findings by Longobardi and Villani (2008) that geology is the major factor affecting baseflow and by Mwakalila et al. (2002) that granitic catchments generate greater baseflow than other catchments.

Further research should test these findings against detailed soil moisture and water table levels. Woolridge et al. (2003) recommend use of soil moisture data as a more information-rich time series than discharge for the calibration of hydrological models in ephemeral catchments. However, in southern Zimbabwe such data is far scarcer than the limited rainfall and discharge data sets. Given the high variability in soil moisture content that is typical of semi-arid catchments (Gómez-Plaza et al., 2001) the utility of this measure for calibration is likely to be limited to experimental catchments.

The clear negative relationship between catchment size and runoff coefficient (Figure 5.2), indicates that smaller catchments are more efficient in converting rainfall into runoff. This is not related to rainfall, as the smaller catchments (Table 5.1) are evenly-distributed between higher rainfall and lower rainfall areas (Table 5.3). Such a scale effect was also observed in studies of micro-catchments (below 1.5 km^2) in both humid to sub-humid environments (Cerdan et al., 2004; Castro et al., 1999; van de Giesen et al., 2000; Didszun and Uhlenbrook, 2008) and arid to semi-arid environments (Cantón et al., 2001; Joel et al., 2002). At the micro-catchment scale, factors such as spatially variable infiltration and the length of the slope (distance of overland flow before entry of runoff into a stream) control what proportion of site runoff is discharged from that scale as runoff and what proportion is redistributed to become soil moisture (van de Giesen et al., 2000). At the meso-catchment scale of this study, processes after runoff generation must operate upon the streams in order to result in the scale relationship observed, for example the re-infiltration of water from streams into groundwater or infiltration or evaporation losses in wetlands. Because of this scale relationship, upscaling parameters from smaller to large catchments, even nested catchments, is complex and can lead to over-estimation of the effects of a given phenomenon (such as erosion) that is measured at a smaller scale when upscaled to a larger catchment or basin.

5.5.4. *Implications for river basin management*

The HBVx model could be used to predict flows in some ungauged basins, although application outside the study area would require calibration to the new basin. As both the regionalisation and blind regionalisation have shown, it cannot be used reliably to predict flows in ephemeral or more arid basins.

The finding that the more degraded catchments are associated with faster flows (likely overland flow) has important implications for flood response management. From this perspective, catchment restoration could potentially play an important role in flood management, by decreasing faster flows and decreasing overland flow.

There are also important implications for setting up hydroclimatic networks in the semi-arid environment. Establishing a network of rainfall stations that has a sufficient density to ensure that aggregated rainfall at catchment level is representative is likely to be costly – especially considering that such a network would only record a few events per year in the more arid catchments. In cases where budgetary constraints prevent a sufficiently dense rainfall measurement network being established it is also essential to fully exploit remote sensing. Observations from an insufficient number of measuring stations will give uncertain results, leading to unrealistic management decisions – unless the level of uncertainty is clearly understood. In such cases the PUB approach represents a better scientific basis for hydrological study and for river basin management. However, the best results are likely to be obtained from remote sensing combined with *in situ* observatations.

The parameter sets that performed best in the regionalisation exercise are suggestive of slow infiltration and percolation with moderate to fast overland flow, which is in line with the general process understanding of such catchments (e.g. Mugabe et al., 2007). These processes appear more extreme in the more degraded catchments. This suggests that rainwater utilisation could be improve through (i) in-field soil water conservation techniques that increase the rate of infiltration and percolation, such as mulching (Mupangwa et al., 2007), and (ii) micro-catchment or runoff farming and supplementary irrigation (Ncube et al., 2009) to capture overland flow from areas adjacent to fields. This is particularly important in the degraded catchments that have faster overland flow. Faster, and possibly greater, sub-surface contribution to streamflow is expected from catchments underlain by granitic rocks.

5.6. Conclusions

Analysis of flow duration curves allowed separation of the catchments with flow over two-thirds of the time (Group C) from the catchments with flow less than 60 % of the time (Groups A and B). This distinction could not be demonstrated statistically. However, this could be due to the small sample size of Group A or could suggest a limitation to agro-ecological region classification as a predictor of catchment behaviour.

Modelling at a daily time step using data from the national hydrological and meteorological networks is not practical for hydrological modelling, due to sparse coverage and the errors discussed above.The two causes are inter-linked and have implications for hydroclimatic network design. Even at a coarser time scale (dekad or monthly), none of the regionalised parameter sets could produce model performance better than C_{NS}=0.3 for half of the catchments studied. The best-performing parameter sets produced mainly negative volume errors (dV_d), which are conservative for water resource modelling and water allocation but problematic for flood prediction. The regionalisation of the HBVx model carried out in this study of the Mzingwane Catchment is thus only partially successful. This could perhaps be improved by incorporating the possibility for negative volume into the lower storage box SL in order to simulate groundwater levels falling below the riverbed (Líden, 2000). This would be an alternative to the approach followed in this study of restricting the flow coefficient from that box (K_2) to very low values.

Hydrological processes in the meso-catchments studied are dominated by slow infiltration and percolation with moderate to fast overland flow, which suggests that rainfed farming would benefit from conservation farming methods that optimise infiltration and percolation, and micro-catchment harvesting of overland flow.

The best performance was obtained at the dekad and monthly time steps, although evaluation by blind regionalisation was limited by data availability. The longer time steps (rather than daily) suggest that the model offers more to longer term water resources (and water allocation) planning than to process hydrology. This also presents an opportunity for coupling HBVx with the spreadsheet-based water balance model WAFLEX, for which an alluvial groundwater module has been developed that performs best at the dekad time step (Love et al., 2010c). The HBVx model thus does offer some limited assistance to river basin planning in semi-arid basins, particularly for predicting flows in ungauged catchments at longer time steps. However, the model is unreliable for more ephemeral and drier catchments.

Ultimately, without more reliable rainfall and runoff data with longer time series, regionalisation in semi-arid ephemeral catchments will remain highly challenging. Data availability, at the appropriate scale and the appropriate density, remains a major challenge in the semi-arid zone, both to river basin management and to accurate regionalisation between catchments. Remotely-sensed rainfall data, such as Tropical Rainfall Measuring Mission (TRMM), can assist, although there are often problems with the magnitude of the estimations (e.g. Hughes, 2006; Arias-Hidalgo et al., 2012). Thus, combining remotely-sensed data with available ground stations may be the way forward.

5.7. Supplementary Material

Table 5.14. Parameter sets resulting from calibration (Table 5.9). MAXBAS was set at 1.0.

	Time step	FC (mm)	LP (-)	β (-)	PE RC (mm d⁻¹)	UZL (mm)	K0 (d⁻¹)	K1 (d⁻¹)	K2 (d⁻¹)
Limits		10- 150	0.70	1- 5	0.1 - 2.5	10 - 100	0.5 - 1.0	0.05- 0.30	0.0001 - 0.0050
B26s	Daily	150.00	0.70	5.00	1.35	72.85	0.999	0.050	0.0001
B78	Daily	150.00	0.70	1.00	2.50	100.00	0.500	0.085	0.0001
B90	Daily	150.00	0.70	4.89	2.50	87.28	0.587	0.050	0.0008
UBN	Daily	150.00	0.70	5.00	2.50	24.93	0.998	0.050	0.0001
MSH	Daily	10.00	0.70	1.00	0.10	23.63	0.896	0.050	0.0001
M27	Daily	150.00	0.70	5.00	2.50	72.42	0.998	0.050	0.0001
B30	Daily	No daily data							
B39	Daily	125.21	0.70	5.00	0.10	99.12	0.617	0.273	0.0050
B60s	Daily	150.00	0.70	1.53	2.50	100.0	0.500	0.057	0.0001
B61s	Daily	150.00	0.70	1.61	2.50	100.0	0.500	0.050	0.0001
B74	Daily	150.00	0.70	5.00	0.13	100.0	0.500	0.245	0.0050
B80	Daily	150.00	0.70	1.54	2.50	100.0	0.500	0.050	0.0001
B11	Daily	No daily data							
B15	Daily	150.00	0.70	1.64	2.50	100.0	0.500	0.055	0.0001
B83	Daily	150.00	0.70	2.07	2.50	100.0	0.500	0.050	0.0001
B26s	Dekad	150.00	0.70	5.00	1.35	72.86	1.000	0.050	0.0001
B78	Dekad	150.00	0.70	1.84	0.10	98.48	0.720	0.300	0.0050
B90	Dekad	150.00	0.70	1.00	0.19	99.19	0.599	0.173	0.0004
UBN	Dekad	102.66	0.70	1.44	1.13	97.46	0.795	0.050	0.0001
MSH	Dekad	10.00	0.70	1.00	0.67	77.49	0.541	0.300	0.0001
M27	Dekad	55.08	0.70	2.82	1.10	83.07	0.654	0.124	0.0002
B30	Dekad	No dekad data							
B39	Dekad	10.00	0.70	1.00	0.10	10.00	1.000	0.257	0.0050
B60s	Dekad	110.98	0.70	1.16	0.67	98.42	0.624	0.300	0.0003
B61s	Dekad	12.45	0.70	1.03	2.07	90.23	0.802	0.253	0.0002
B74	Dekad	12.36	0.70	5.00	0.27	10.37	0.924	0.300	0.0050
B80	Dekad	128.21	0.70	1.64	0.24	26.06	0.551	0.300	0.0001
B11	Dekad	No dekad data							
B15	Dekad	40.55	0.70	1.52	0.55	10.00	0.500	0.300	0.0001
B83	Dekad	85.45	0.70	1.82	0.10	93.53	0.630	0.300	0.0001

	Time step	FC (mm)	LP (-)	β (-)	PE RC (mm d^{-1})	UZL (mm)	K0 (d^{-1})	K1 (d^{-1})	K2 (d^{-1})
Limits		10-150	0.70	1-5	0.1 - 2.5	10 - 100	0.5 - 1.0	0.05-0.30	0.0001 - 0.0050
groupABC	Dekad	42.01	0.70	1.00	0.69	17.18	1.000	0.300	0.0001
B26s	Monthly	150.00	0.70	1.00	0.82	88.94	0.738	0.300	0.0001
B78	Monthly	150.00	0.70	1.00	0.63	92.65	0.772	0.300	0.0001
B90	Monthly	12.20	0.70	1.02	0.26	99.88	0.634	0.256	0.0021
UBN	Monthly	Insufficient data							
MSH	Monthly			Insufficient data					
M27	Monthly			Insufficient data					
B30	Monthly	10.00	0.70	1.00	0.14	96.17	0.524	0.300	0.0050
B39	Monthly	10.00	0.70	1.00	0.10	99.49	0.527	0.300	0.0050
B60s	Monthly	57.29	0.70	1.00	0.42	98.65	0.521	0.300	0.0005
B61s	Monthly	10.00	0.70	1.00	0.82	84.48	0.518	0.300	0.0007
B74	Monthly	10.00	0.70	1.00	0.11	10.00	0.733	0.300	0.0001
B80	Monthly	60.42	0.70	1.23	0.24	99.15	0.738	0.300	0.0001
B11	Monthly	10.00	0.70	1.00	0.41	10.00	1.000	0.300	0.0009
B15	Monthly	10.00	0.70	1.00	0.63	87.55	0.551	0.300	0.0001
B83	Monthly	150.00	0.70	1.19	0.34	86.66	0.502	0.300	0.0001
groupA	Monthly	150.00	0.70	1.00	0.52	96.43	0.538	0.300	0.0001
groupAB	Monthly	10.00	0.70	1.00	0.43	10.00	1.000	0.300	0.0001
groupB	Monthly	10.00	0.70	1.00	0.38	10.00	1.000	0.300	0.0001
groupC	Monthly	10.00	0.70	1.00	0.53	10.00	1.000	0.300	0.0001

6. Targeting the under-valued resource: an evaluation of the water supply potential of small sand rivers in the northern Limpopo Basin [*]

6.1. Abstract

The alluvial aquifers that form the beds of sand rivers are perennial in semi-arid regions of ephemeral rivers, largely protected from evaporation and normally of good quality. The northern Limpopo Basin has erratic and unreliable rainfall and very low mean annual runoff. Its alluvial aquifers thus present a sustainable alternative to surface water use. However, few catchments are gauged, alluvial aquifers have been mapped in even fewer catchments and hydrogeological information is only available for a small number of sites.

The water supply potential of a case study alluvial aquifer was evaluated using field observations and the finite difference groundwater flow model, MODFLOW. The behaviour of the aquifer under higher seepage, and climate change and development scenarios was also modelled. This showed that alluvial aquifers of this scale are suitable for use for domestic and livestock water supply and the irrigation of small gardens (sub-hectare to several hectares per km of river reach). The major effect of climate change on the aquifer is likely to be increased drawdown of storage (5 to 15 %) due to decreased river flows. A remote sensing approach was used to identify and map 1,835 km of alluvial aquifers in the northern Limpopo Basin. Modelling showed that alluvial aquifers hosted on North Marginal Zone bedrock, Karoo basalt and Chilimanzi granites should have the least seepage, which could support 2,550 ha of irrigation year-round (excluding 7,480 ha along the lower Mzingwane river and the alluvial aquifers of the Limpopo river itself). The methods used in this study can be applied in other semi-arid areas to allow rapid targeting of sand rivers for field investigation and potential water supply development.

6.2. Introduction

The sandy beds of rivers world-wide host alluvial aquifers. Because of their shallow depth and their vicinity to the streambed, alluvial aquifers have an intimate relationship with surface flow, which is the main source of aquifer recharge. Indeed it can be argued that groundwater flow in alluvial aquifers is an extension of surface flow (Mansell and Hussey, 2005; Love et al., 2010c). In arid and some semi-arid areas, alluvial aquifer recharge may occur only after high discharge peaks from heavy rainfall events (Lange and Leibundgut, 2003; Lange, 2005; Matter et al., 2005; Benito et al., 2010) and full recharge normally occurs early in the rainy season (Owen and Dahlin, 2005). No surface flow occurs until the channel aquifer is saturated (Nord, 1985).

---

[*] Based on: Love, D.; Owen, R.J.S.; Uhlenbrook, S.; van der Zaag, P. *submitted*. Targeting the under-valued resource: an evaluation of the water supply potential of small sand rivers in the northern Limpopo Basin. Submitted to *Water Resources Management*.

The water supply potential of large alluvial aquifers, which are seen as good sources for irrigation water, has been studied extensively (Owen and Dahlin, 2005; Moyce *et al.*, 2006; Raju *et al.*, 2006; De Leon *et al.*, 2009). Little scientific work has been done on small alluvial aquifers – here understood to refer to aquifers on sand rivers draining a meso-catchment (catchment area of approximately $10^1 - 10^3$ km^2; Blöschl and Sivapalan, 1995). Furthermore, water resources development plans tend to similarly focus on large scale (alluvial) aquifers (e.g. MCC, 2009). This is despite the fact that communities in Africa have abstracted water from such sand rivers for generations, often through artisanal wells (De Hamer *et al.*, 2008; Agyare *et al.*, 2009) or riverine dugouts (Harrington *et al.*, 2008; Ofosu *et al.*, 2010). Whilst these aquifers will have lower potential storage than larger ones, small alluvial aquifers may be easier to access for poor rural communities. This is because a smaller head difference between the riverbed and the bank can allow for cheaper or manual pumps. In addition, smaller streams are likely to be more widely distributed than the major channels. Thus accessing the alluvial aquifers of small sand rivers for irrigation represents a development possibility for smallholder farmers, and therefore an important opportunity for improving food production. However, little knowledge is available on the hydrogeological characteristics of small alluvial aquifers.

In the Mzingwane Catchment (the Zimbabwean part of the Limpopo Basin) work to date on alluvial aquifers has focussed on the major tributaries of the Limpopo River, such as the Shashani and Thuli Rivers (Mansell and Hussey, 2005) and the Mzingwane River (Owen and Dahlin, 2005; Moyce *et al.*, 2006; Masvopo *et al.*, 2008; Love *et al.*, 2010c). This study focuses on the alluvial aquifers of the small sand rivers that are tributaries to these major rivers. The objectives of the study are: (i) to evaluate the water supply potential of a case study of a meso-scale alluvial aquifer, using field observations and water balance modelling; (ii) to determine the behaviour of the case study aquifer under different hydrogeological conditions, such as higher seepage; (iii) to predict the effect on the aquifer of climate change and development scenarios such as upstream releases and increased upstream demand; and (iv) to regionalise a desktop evaluation of the water supply potential of the alluvial aquifers in small sand rivers for the purpose of targeting field investigations of possible development sites.

6.3. Methods

6.3.1. Case study site

The case study site is in the Mushawe river, Mwenezi District (see Figure 1.1 for location). This is an area with low mean annual runoff and frequently saline groundwater (Hoko, 2005). Alluvial groundwater, available year-round and fresher than deep groundwater (Moyce *et al.*, 2006) is thus a strategic alternative. The Mushawe river is a right bank tributary of the middle Mwenezi river. At Maranda, the Mushawe is a sand river with a catchment of 220 km^2 (Figure 6.1) and a mean annual runoff of 50 mm a^{-1}. The sand river forms an alluvial aquifer, blocked by the concrete foundation of a bridge that functions as a barrier to sub-surface flow of water, as well to the downstream movement of sand. This is similar to the effect of constructed sand storage dams (Aerts *et al.*, 2007) or gabion weirs (Mansell and Hussey, 2005). The site supplies water to the adjacent Maranda No 1 Business Centre (approximately 5 m^{3}day^{-1}). It is a fairly typical example of a lowland tributary in the Mzingwane Catchment. The mean annual rainfalls for the nearest reference rainfall stations are 472 mm a^{-1} at Mberengwa (1987-2000) and 421 mm a^{-1} at West Nicholson (1962 – 2008; Love et al., 2010b). The alluvial groundwater in the Mushawe river is fresh, not saline (Mandiziba, 2008), compared to groundwater in the bedrock (222 to 9,800 µS cm^{-1}, Hoko, 2005). Upstream of No 1, the river is 35 km long and the catchment is composed of fields (32.4 %), medium to dense woodland (13.2 %), mixed grassland and woodland (48.4 %), rocky hills (5.6 %) and wetland (0.4 %), see Figure 6. 1. The population of the catchment can be estimated at around 11,200, based on available figures for the municipal wards (Parliament of Zimbabwe, 2008). The catchment is underlain solely by granitoids (mainly tonalities and trondhjemites) of the Limpopo Belt North Marginal Zone, aged 2.72 - 2.52 Ga (Rollinson and Blenkisop, 1995; Chinoda et al., 2009). These rocks have been found to be resistant to weathering (Butterworth et al., 1999).

6.3.2. Data collection

The surface of the riverbed, channel width and slope were surveyed in the field and the depth of sand determined by physical probing with a steel probe. Composite samples of alluvial material were collected from various depths of the aquifer. Grain size distribution was determined by the sieve shaker method. The sieves used were US standard sieves with sizes 4000, 2800, 2000, 1000, 500, 250, 180, 125 and 32 µm. Porosity (n) was then derived indirectly using the coefficient of grain uniformity (Vukovic and Soro, 1992):

$$n = 0.255(1 + 0.83^U) \qquad U = \frac{d_{60}}{d_{10}} \qquad (6.1)$$

Where n = porosity (-), U = coefficient of grain uniformity (-), d_{60} = sieve size for which 60 % of the sample passed (mm) , d_{10} = sieve size for which 10 % of the sample passed (mm).

Specific yield was determined from the volume of water that drained under gravity from a known volume of saturated aquifer material:

$$S_y = \frac{V_{wd}}{V_{tot}} \qquad (6.2)$$

Where S_y = specific yield (-), V_{wd} = volume of water drained under gravity (cm^3), V_{tot} = total volume of saturated sand (cm^3).

Hydraulic conductivity was determined by using the permeameter method: a 0.06 m^3 bucket with a 25 mm outlet at the base was completely filled with aquifer material and a constant head maintained by continuous inflow. The time taken for water flowing out of the outlet to fill a 0.05 m^3 bucket was recorded and the hydraulic conductivity computed as follows:

$$K = \frac{V}{tA} \qquad (6.3)$$

Where K = hydraulic conductivity (m d^{-1}), V = volume of water flowing out of the outlet (m^3), t = time (d), A = cross-sectional area of bucket (m^2).

Seven piezometers (Figure 6. 2) were driven through the aquifer to the top of the bedrock, placed in pairs at four locations upstream of the bridge (Figure 6. 3), which forms the downstream edge of the study site (see Figure 6. 1). Discharge was measured using a limnigraph located on the bridge and the river cross-section at the bridge, which was surveyed. Daily observations were made of the water level in each piezometer and at the limnigraph. Rainfall was measured daily, using 17 rain gauges, spread throughout the catchment (see Figure 6. 1), and catchment rainfall was calculated using Thiessen polygons.

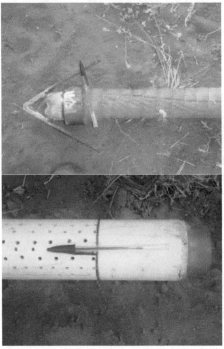

Figure 6.2. Detail of piezometers used: (left) steel tip and geomembrane-wrapped well point, (right) outer casing.

Figure 6.3. Aerial view of the study site during the field work period (5 January 2007); showing the riverbed, location of instrumentation and cells used for storage computations. L = limnigraph, P1-P8 = piezometers no. 1 – no. 8, A = abstraction point.

6.3.3. *Groundwater storage and usage*

The riverbed was divided into seven cells, each allocated to a piezometer (Figure 6.3). For each pair of adjacent cell, a cross-section was determined from the topographic survey of the river channel, the physical probing and the depths of the piezometers.

The volume of water stored in aquifer on a given date was determined from the specific yield, the daily piezometer readings, the depth measurements and the riverbed surface area:

$$V_{pot} = S_y \sum_{i=Pz1}^{i=Pz8} \frac{A_i l_i}{2} \tag{6.4}$$

$$V_{gw} = S_y \sum_{i=Pz1}^{i=Pz8} \left(\frac{h_i}{h_{pz}} \times \frac{A_i l_i}{2} \right) \tag{6.5}$$

Where V_{gw} = volume of stored groundwater (m^3), V_{pot} = potential volume of stored groundwater (m^3), A_i = cross-sectional area of riverbed cell containing piezometer i (m^2), l_i = length of cell containing piezometer i (m), h_i = height of watertable. in piezometer i (m), h_{pz} = length of piezometer i (m), S_y = specific yield (-).

Water usage by the business centre was considered on the basis of figures obtained from the business centre pump operative (5 m^3d^{-1}). Water usage for year-round irrigation are based upon based on 10,000 $m^3ha^{-1}a^{-1}$ (Ministry of Local Government Rural and Urban Development, 1996), equivalent to 27.4 m^3d^{-1}. This daily figure was used to estimate irrigation water demand for dry season irrigation (April to September) and for supplementary irrigation during dry spells. A dry spell is considered to be a period of 5 days or longer with rainfall of less than 1.0 mmd^{-1}. For the purpose of supplementary irrigation, a maize crop was considered, with water required for 180 d from planting until harvest (FAO, 2013b). Supplementary irrigation of rainfed agriculture is considered a major potential intervention for improving the resilience of rainfed agriculture (Rockström *et al.*, 2003; Falkenmark and Rockström, 2004; Magombeyi and Taigbenu, 2008).

6.3.4. *Groundwater flow model*

A groundwater flow model of the aquifer was prepared using the finite difference, distributed MODFLOW model, in the Visual MODFLOW 4.3 interface, which is widely used for three-dimensional stream-aquifer modelling (e.g. Scibek and Allen, 2006, Prundeda *et al.*, 2010), including of alluvial aquifers (Rodríguez *et al.*, 2006). The conceptual design of the model is shown in Figure 6. 4 and the data used to set up the model is shown in Table 6.1. The model was run using MODFLOW 2000 and ZoneBudget (Harbaugh, 2005).

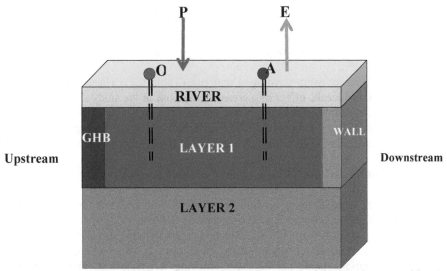

Figure 6.4. Conceptual design of the MODFLOW model of the Mushawe aquifer. P = precipitation recharge, ET = evapotranspiration, O = observation well, A = pumping well, GHB = general head boundary (groundwater flow from upstream), WALL = bridge blocking groundwater flow downstream, RIVER = Mushawe River (ephemeral), LAYER 1 = alluvial aquifer, LAYER 2 = tonalite bedrock.

Table 6.1. Sources of data used in MODFLOW

Model component	Data Requirement	Data Source
Aquifer boundary	Positions	Surveyed
RIVER boundary	Conductance	

Model component	Data Requirement	Data Source
	River stage	Limnigraph observations
General Head Boundary	Boundary head	Piezometer 7 & 8 observations
WALL boundary	Bridge treated as impermeable wall	Field observations
Recharge	Rainfall	Rain gauge observations
Evapotranspiration	Evapotranspiration	CROPWAT (Masvingo)
Aquifer properties	Hydraulic conductivity	Field experiments
Basement properties	Hydraulic conductivity	Basement hydraulic conductivity was considered to be very low based on field observations and the literature[*]
Pumping well	Abstraction	Pump operator's records
Observation wells	Head	Piezometer 1- 5 observations (Piezometer 2 was not used due to its proximity to the pumping well leading to subdaily variability)

[*] e.g. Butterwork *et al.* (1999). Compare also to values for granitoids of 2.16 $\times 10^{-7}$ m d^{-1} (Davis, 1969) or 5.18 $\times 10^{-7}$ m day^{-1} (Abelin *et al.*, 1991)

The model was validated against data from the observation wells, using the Nash-Sutcliffe Coefficient (equation 4.10). A simple sensitivity analysis was carried out for the validated model. The 10 % elasticity index (equation 4.12) was used.

A series of scenarios were developed and run (Table 6.2). These scenarios were developed to consider the effect of increased water use at the study site, or upstream of the study site and the effects of climate change, the loss of sand bed material and other possible changes at the study site. Different basement geologies were also modelled, to consider the dynamics of a similar alluvial aquifers other sites.

6.3.5. Identification and classification of alluvial aquifers

Geo-referenced LandSat imagery was used to identify alluvial aquifers. False colour composites of bands 3, 4 and 5 were prepared of LandSat scenes path 169 row 075 (1 June 2005), path 170 row 175 (3 December 2000) and path 171 row 075 (11 January 2001). Alluvial channels sands stand out as white and dense actively growing vegetation stands out as green, making it possible to identify alluvial channel deposits (Moyce *et al.*, 2006). The identified alluvial aquifers were then classified on the basis of basement geology (using mapping by Chinoda *et al.*, 2009) and erosion surface (Twidale, 1988). The occurrence of approximately half of the aquifers identified and their bedrock geology were verified by field visits.

6.4. Results and Discussion

6.4.1. Field data

The hydrogeological characteristics of the aquifer are presented in Table 6.3 and show properties that are reasonable for an aquifer composed of fine sand. The porosity value is somewhat higher than the 30 % derived by Moyce *et al.* (2006) and the 35 % derived by Owen (1991) for the Mzingwane River and also by Nord (1985) for rivers in neighbouring Botswana. The specific yield is within the expected range for fine sand (Johnson, 1967), which fits with the observed grain size distribution. It is also within the range for alluvial aquifers reported elsewhere (e.g. Barlow *et al.*, 2003; Rodríguez *et al.*, 2006), and close to that reported by Nord (1985) and Owen (1991) from neighbouring river basins. The hydraulic conductivity is on the border between that expected for fine sand and that expected for silt (Bear, 1972), which also fits with the grain size distribution. Note that some authors assume that specific yield is equal to the aquifer effective porosity and thence to the total porosity - see for example the discussion of this problem in Laslett and Davis (1998). In this case, such an assumption would result in a 300 % increase in the volume of stored water calculated.

Table 6.3. Hydrogeological characteristics of the Mushawe alluvial aquifer

Channel slope	0.17 %
Width of river at limnigraph	50.0 m
Depth of alluvial aquifer material (range)	1.60 to 2.45 m
Depth of alluvial aquifer material (average)	2.10 m
Average cross-sectional area of aquifer	106 m^2
Grain size distribution	Fine sand to silt 54 %, sand 43 %, clay 2 %, coarse sand to gravel 2 %
Porosity	43 %
Specific yield	14.4 %
Hydraulic conductivity	26.8 m d^{-1}

When comparing discharge and depth to watertable (Figure 6.5), it can be seen that a small volume of water flowed out of the aquifer at the bridge throughout the year – discharge in March to April 2008 was non-zero, but too low to reflect on the graph. There were no other low flows, and four floods during the period of observation. The occurrence of floods coincided reasonably with periods of rainfall of at least 30 mm over 3 days (see supplementary material).

During the rainy season the watertable initially rose (early December 2007), and then reached the surface allowing for discharge events. The absence of a significantly falling watertable. during the dry season indicates minimal seepage. This is in contrast to aquifers further west (see Figure 1.1) where seepage was found to be a major flux (Love et al., 2007; De Hamer et al. 2008). This is probably related to differences in the bedrock: Mushawe overlying North Marginal Zone tonalites, as opposed to older gneisses at the other sites.

The discharge hydrograph does not present conventional rise and recession behaviour. During the four peak surface flow events of 18-20 December 2007, 26-29 December 2007, 7-13 January 2008 and 26-27 January 2008, the surface water and groundwater hydrographs rise and fall together within the timestep of the observations. This could be explained by considering the aquifer and river as a single unit: water flowing towards the study site, whether surface flow in the river upstream or overland flow, must first saturate the aquifer before surface flow occurs (excluding the small discharge from the aquifer at the bridge). This effect can be seen in early December 2007, where the groundwater hydrograph rises – and can be seen to relate to upstream rainfall events – but the surface water hydrograph does not respond until late December 2007.

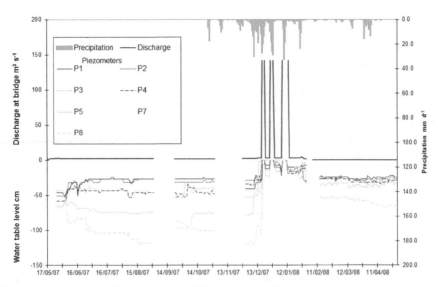

Figure 6.5. Observed discharge, rainfall and watertable. level in the study area. Discharge readings go to a maximum of 141.6 m³s⁻¹, above which the limnigraph and bridge were overtopped and no measurements could be made.

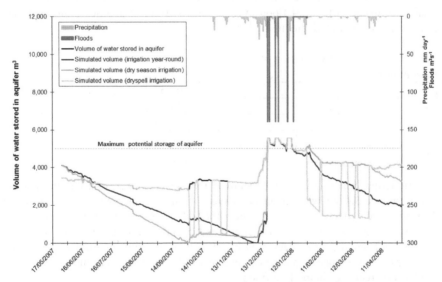

Figure 6.6. Observed storage levels in the Mushawe alluvial aquifer, with rainfall and flood events (flows over 2.5 m³s⁻¹) for reference. The simulated volume shows a scenario of abstraction of sufficient water for year-round irrigation of a 0.9 ha garden, a scenario of dry season irrigation of a 1.2 ha garden, and a scenario of supplementary irrigation of 102 ha of maize to bridge dry spells. The simulations start from the aquifer volume recorded at the end of the study period (4,100 m³ at the end of April).

6.4.2.   Groundwater storage and usage

The aquifer was fully saturated by December 2007 (5,425 m^3) but drops to 4,000 m^3 as the 2008 dry season begins (Figure 6.6). The volume of water stored in the aquifer at the end of the 2008 dry season is higher than that at the end of the 2007 dry season, probably due to the rainy season of the year preceding the 2007 dry season having less rainfall, during an El Niño event (Logan et al., 2008). Each of the three flood events resulted in the (temporary) filling of the aquifer's full potential storage. Simulation of irrigation demand shows that, in addition to meeting the current water demand of the business centre, sufficient water can be supplied from the aquifer reach studied for year-round irrigation of 0.7 ha, or 0.9 ha if the demand for the business centre is excluded. If irrigation demand is only during the dry season, then a garden of 1.0 ha can be supplied, or 1.2 ha if the demand for the business centre is excluded. Supplementary irrigation to bridge the dry spells observed during the field season could theoretically be supplied for 102 ha of maize.

6.4.3.   Groundwater flow modelling

The model performed well when validated against the piezometer heads observed in the field (Table 6.4). The sensitivity analysis (Table 6.5) showed that the model is not very sensitive to the abstraction well (probably as the volumes remain very small) nor to the volume of flood events. Sensitivity to hydraulic conductivity and evaporation extinction depth are higher.

Table 6.4. Results of validation of MODFLOW model

Piezometer	C_{NS}
P1	0.655
P3	0.613
P4	0.608
P5	0.410
Overall	0.567

The modelling results (Table 6.6) showed that an area of 1.0 ha can be irrigated year-round or 3.0 ha with managed releases from an upstream dam. If upstream demand merely decreases the volume of flood events, there is no major effect. However when the low flows are eliminated by upstream demand, the volume of infiltration from the river decreases by 49%. Construction of sand dams, gabion weirs or other sub-surface barriers upstream would not affect the study site as all of these would recharge in the first flood.

Considering climate change, increases in evapotranspiration or decreases in rainfall, on their own, had no significant effect on the model. However, the decrease in river flow volumes had an immediate effect: a 25% decrease in flow resulting in a 5% drawdown of storage and a 35% decrease in flow resulting in a 14% drawdown of storage. This rises to 15% drawdown when the decrease in river flow is coupled with decrease in rainfall and increase in evapotranspiration.

Simulation of a massive basalt basement, similar to that occurring elsewhere in the Limpopo Basin, was found to result in seepage that was higher than that of unweathered granite, but still small enough (89 m^3a^{-1}) to have no significant effect on the aquifer water balance. Sandstone basement showed variable seepage, with lower levels of seepage having no significant effect but with higher levels resulting in a 24% drawdown of storage. Substantially higher seepage was found to occur on African erosion surface granites (168 to 928 m^3a^{-1}) and fractured granites (729 to 1570 m^3a^{-1}), leading to substantial increases in storage drawdown – an increase of 205% for the fractured granites.

6.4.4.   Targeting: identification and classification of alluvial aquifers

Remote sensing of the Mzingwane Catchment identified 1,835 km of alluvial aquifers (Figure 6.7). Modelling of the effect of different bedrock lithologies and erosion surfaces (see Table 6.6), as well as previous studies (Butterworth *et al.*, 1999; De Hamer *et al.*, 2008; Larsen *et al.*, 2002; Love *et al.*, 2010b; Owen and Madari, 2009), were used to indicate likely comparative levels of seepage. Alluvial aquifers hosted on North Marginal Zone bedrock, Karoo basalt and Chilimanzi granites are suggested to have lower levels of seepage. Alluvial aquifers hosted on older gneisses are suggested to have higher levels of seepage, and it is likely that Beitbridge gneisses behave in a similar fashion. The following bedrock types are either heterogeneous in permeability or there is insufficient information available: Achaean greenstone belts, Karoo and Cretaceous sediments and Karoo Felsic volcanics. This indicates approximately 780 km of alluvial aquifers with likely minimal seepage (see Figure 6.7).

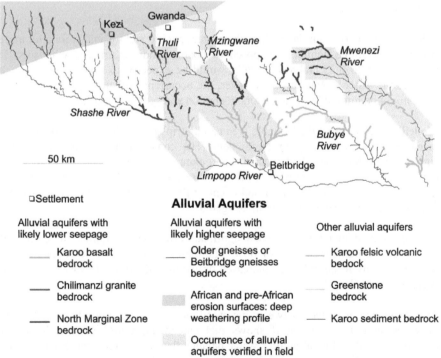

Figure 6.7. Alluvial aquifers of the Mzingwane Catchment, identified from satellite imagery and classified on the basis of likely seepage.

Scale also imposes limitations on aquifer viability: smaller catchments tend to have shallower river sands, and where sand depth is less than evaporation extinction depth (0.9 m), storage potential is likely to be negligible (Love et al., 2007). However, such catchments are unlikely to be included in the remote sensing assessment, as they are likely to be too small to have been identified.

6.5. Discussion

The maximum storage of the aquifer studied is saturated during all flood events. Only around half of the groundwater remains stored in the aquifer during the year, but this residual volume is nevertheless sufficient water to serve the water supply to No 1 Business Centre and potentially also a small irrigated area. There are approximately 23 km of alluvial aquifer upstream of the 0.6 km reach that was studied. Should these 23 km hold the same potential as the reach studied, the catchment storage would be approximately 21,000 m^3, or 1.9 % of mean annual runoff, which is comparable with the lower end of findings from Kenya reported by Aerts et al. (2007). This means that up to 35 ha could be irrigated year-round or 46 ha in the dry season only, or around 3,900 ha for supplementary irrigation of rainfed maize. However, these reaches are upstream of the area studied and so are likely to have shallower sand fills, decreasing the storage potential. Furthermore the bridge provides favourable dry season conditions at the study site, by functioning as a sub-surface barrier and minimising downstream groundwater flow. There is only one similar site in the study area, Chengwe (upstream, see Figure 6.1) and no similar sites downstream of the study area along the Mushawe River – although sand dams could be constructed. Furthermore, whilst the water required for 3,900 ha of supplementary irrigation could be supplied, the infrastructure costs probably outweigh the benefits.

Alluvial aquifers of a similar scale with greater specific yield would store a lot more water: for example aquifers composed of coarse sands or gravels, with specific yields of 0.25 to 0.35 (Johnson, 1967), would be expected to store twice as much abstractable water as Mushawe. However, the occurrence of such aquifers at meso-scale might be limited by scale factors in erosion and sediment transport dynamics.

Modelling showed that the irrigable area can be tripled if recharge were possible by managed releases from an upstream dam. The construction of sand dams, of a similar size to that controlled by the bridge, along the 23 km of aquifer upstream of the study site is possible in terms of the catchment water balance, as all sand dams would be recharged by the first flood event of the rainy season. However, possible sites for such sand dams were not investigated.

New upstream demand would have a significant effect on the aquifer studied only if it affected the number of days of flow. The major effect of climate change on the aquifer is likely to be indirect: increased drawdown of storage (5 to 15 %) due to decreased river flows.

The difference in effect between increased/more frequent rainfall compared to increased/more frequent runoff shows that catchment runoff is a much greater component of the alluvial aquifer water balance than direct recharge from rainfall. An increase in the frequency of flows has a greater effect than an increase in their volume.

Extensive alluvial aquifers have been shown to exist on all primary and many secondary tributaries of the Limpopo River within the Mzingwane Catchment (Figure 6.7). For around one third of the aquifer reaches, bedrock conditions favour minimal seepage losses, allowing for targeting of field investigations and possible development. These favourable sites comprise some 780 km of alluvial aquifers, which, if of similar groundwater potential to the Mushawe site, could irrigate some 1,150 ha year-round or 1,550 ha during the dry season. However, it is likely that many of these aquifers are considerably deeper – at the depth of 4.5 m reported from the middle Mzingwane river, 2,550 ha of irrigation could be supported year-round or 3,350 ha in the dry season only. The additional capacity can provide for 2,550 households at 0.29 ha per household. This is in addition to 7,480 ha that can be irrigated from aquifers previously mapped along the lower Mzingwane river (Love et al., 2010c) and the alluvial aquifers of the Limpopo river itself, which are already heavily used by South African farmers (Boroto, 2001).

6.6.   Conclusions and Recommendation

Alluvial aquifers of the scale of Mushawe are likely to be suitable for use at the scale of domestic and livestock water supply and the irrigation of small gardens (sub-hectare to several hectares per km of river reach). Such gardens would be best suited for horticulture, to maximise the value obtained from the irrigation water, and are replicable on many river reaches. Alternatively, alluvial aquifers of this scale also offer a considerable potential for supplementary irrigation of rainfed farming (over 100 hectares per km of river reach), reaching a much larger number of smallholder farmers.

The small alluvial aquifers offer the potential for distributed, localised storage, readily accessible to a large number of communities with limited financial resources. There is a growing consensus that small-scale water supply technologies are the most cost-effective (Lasage et al., 2007; Van der Zaag and Gupta, 2008) – and such small-scale technologies are both appropriate for alluvial aquifers in small sand rivers and also more likely to be within the reach of smallholder farming communities.

There is widespread unutilised alluvial aquifer potential in the Mzingwane Catchment, capable of supporting 1,250 to 2,800 ha of irrigated land year-round. A simple bedrock classification can assist in targeting favourable reaches of these sand rivers for further evaluation and possible water supply development. The methods used in this study can be applied in other semi-arid areas to allow rapid targeting of sand rivers for field investigation and potential water supply development.

6.7. Supplementary Material

Table 6.7. Comparison of discharge and cumulative rainfall. Discharge values greater than 0.5 m³d⁻¹ and cumulative rainfall values greater than 30 mm over three days are highlighted.

Date	Rainfall (mm / d)	K3 Rainfall (mm / 3 d)	Discharge (m³ / d)	Date	Rainfall (mm / d)	K3 Rainfall (mm / 3 d)	Discharge (m³ / d)
01/11/2007	0.0	0.0	0.00	01/02/2008	0.0	0.5	0.00
02/11/2007	0.0	0.0	0.00	02/02/2008	0.0	0.2	0.00
03/11/2007	0.0	0.0	0.00	03/02/2008	0.0	0.0	0.00
04/11/2007	0.0	0.0	0.00	04/02/2008	0.0	0.0	0.00
05/11/2007	0.7	0.7	0.00	05/02/2008	0.0	0.0	0.00
06/11/2007	4.9	5.6	0.00	06/02/2008	0.0	0.0	0.00
07/11/2007	10.1	15.7	0.00	07/02/2008	0.0	0.0	0.00
08/11/2007	7.4	22.4	0.00	08/02/2008	0.2	0.2	0.00
09/11/2007	0.7	18.2	0.00	09/02/2008	1.4	1.6	0.00
10/11/2007	0.0	8.1	0.00	10/02/2008	1.5	3.0	0.00
11/11/2007	0.0	0.7	0.00	11/02/2008	0.8	3.7	0.00
12/11/2007	0.0	0.0	0.00	12/02/2008	0.0	2.3	0.00
13/11/2007	0.0	0.0	0.00	13/02/2008	0.0	0.8	0.00
14/11/2007	0.2	0.2	0.00	14/02/2008	0.0	0.0	0.00
15/11/2007	0.0	0.2	0.00	15/02/2008	0.0	0.0	0.00
16/11/2007	0.0	0.2	0.00	16/02/2008	0.0	0.0	0.00
17/11/2007	0.0	0.0	0.00	17/02/2008	0.0	0.0	0.00
18/11/2007	0.0	0.0	0.00	18/02/2008	0.1	0.1	0.00
19/11/2007	0.3	0.3	0.00	19/02/2008	0.0	0.1	0.00
20/11/2007	0.0	0.3	0.00	20/02/2008	0.7	0.7	0.00
21/11/2007	0.0	0.3	0.00	21/02/2008	0.5	1.1	0.00
22/11/2007	0.2	0.2	0.00	22/02/2008	0.0	1.1	0.00
23/11/2007	2.0	2.2	0.00	23/02/2008	0.0	0.5	0.00
24/11/2007	2.5	4.6	0.00	24/02/2008	0.0	0.0	0.00
25/11/2007	4.0	8.4	0.00	25/02/2008	0.0	0.0	0.00
26/11/2007	8.6	15.1	0.00	26/02/2008	0.0	0.0	0.00
27/11/2007	2.4	15.0	0.00	27/02/2008	0.0	0.0	0.00
28/11/2007	3.8	14.8	0.00	28/02/2008	0.0	0.0	0.00
29/11/2007	5.1	11.3	0.00	29/02/2008	0.1	0.1	0.00
30/11/2007	0.0	8.9	0.00	01/03/2008	2.0	2.1	0.00
01/12/2007	0.1	5.2	0.00	02/03/2008	0.0	2.1	0.00
02/12/2007	0.0	0.1	0.00	03/03/2008	0.0	2.0	0.00
03/12/2007	0.5	0.6	0.00	04/03/2008	0.4	0.4	0.00
04/12/2007	1.6	2.1	0.00	05/03/2008	0.4	0.8	0.00
05/12/2007	0.3	2.4	0.00	06/03/2008	1.8	2.6	0.00

Date	Rainfall (mm / d)	K3 Rainfall (mm / 3 d)	Discharge (m³ / d)	Date	Rainfall (mm / d)	K3 Rainfall (mm / 3 d)	Discharge (m³ / d)
06/12/2007	2.3	4.3	0.00	07/03/2008	1.7	3.9	0.00
07/12/2007	3.1	5.8	0.00	08/03/2008	0.7	4.2	0.00
08/12/2007	9.0	14.5	0.00	09/03/2008	0.0	2.4	0.00
09/12/2007	13.5	25.7	0.00	10/03/2008	0.0	0.7	0.00
10/12/2007	30.0	52.6	0.00	11/03/2008	0.0	0.0	0.00
11/12/2007	8.4	52.0	0.00	12/03/2008	0.0	0.0	0.00
12/12/2007	9.7	48.2	0.00	13/03/2008	0.0	0.0	0.00
13/12/2007	10.2	28.4	0.00	14/03/2008	0.0	0.0	0.00
14/12/2007	19.0	38.9	0.00	15/03/2008	1.7	1.7	0.00
15/12/2007	3.3	32.5	0.00	16/03/2008	4.5	6.2	0.00
16/12/2007	5.2	27.5	0.00	17/03/2008	7.3	13.4	0.00
17/12/2007	11.8	20.3	0.00	18/03/2008	0.6	12.3	0.00
18/12/2007	27.4	44.4	56.73	19/03/2008	0.0	7.9	0.00
19/12/2007	19.3	58.6	56.73	20/03/2008	0.0	0.6	0.00
20/12/2007	13.1	59.9	56.73	21/03/2008	0.0	0.0	0.00
21/12/2007	5.1	37.6	0.00	22/03/2008	0.0	0.0	0.00
22/12/2007	0.9	19.2	0.00	23/03/2008	0.0	0.0	0.00
23/12/2007	1.3	7.4	0.00	24/03/2008	0.0	0.0	0.00
24/12/2007	2.7	5.0	0.00	25/03/2008	0.0	0.0	0.00
25/12/2007	1.3	5.4	0.00	26/03/2008	0.0	0.0	0.00
26/12/2007	9.5	13.5	56.73	27/03/2008	0.0	0.0	0.00
27/12/2007	18.4	29.1	56.73	28/03/2008	0.3	0.3	0.00
28/12/2007	25.7	53.5	56.73	29/03/2008	15.7	16.0	0.00
29/12/2007	28.8	72.9	56.73	30/03/2008	2.0	17.9	0.00
30/12/2007	1.2	55.7	0.00	31/03/2008	5.6	23.2	0.00
31/12/2007	1.3	31.3	0.00	01/04/2008	3.9	11.4	0.00
01/01/2008	0.7	3.2	0.00	02/04/2008	0.8	10.2	0.00
02/01/2008	0.0	2.0	0.00	03/04/2008	8.0	12.7	0.00
03/01/2008	0.1	0.8	0.00	04/04/2008	13.4	22.1	0.00
04/01/2008	0.0	0.1	0.00	05/04/2008	4.5	25.9	0.00
05/01/2008	0.1	0.2	0.00	06/04/2008	0.0	17.9	0.00
06/01/2008	0.1	0.2	0.00	07/04/2008	0.0	4.5	0.00
07/01/2008	0.5	0.7	56.73	08/04/2008	0.0	0.0	0.00
08/01/2008	7.6	8.2	56.73	09/04/2008	0.7	0.7	0.00
09/01/2008	32.6	40.7	56.73	10/04/2008	0.8	1.5	0.00
10/01/2008	18.6	58.8	56.73	11/04/2008	0.0	1.5	0.00
11/01/2008	5.3	56.5	56.73	12/04/2008	0.0	0.8	0.00
12/01/2008	0.7	24.5	41.23	13/04/2008	0.0	0.0	0.00
13/01/2008	0.0	6.0	39.71	14/04/2008	0.2	0.2	0.00
14/01/2008	1.4	2.1	0.00	15/04/2008	0.2	0.4	0.00

Date	Rainfall (mm / d)	K3 Rainfall (mm / 3 d)	Discharge (m³ / d)	Date	Rainfall (mm / d)	K3 Rainfall (mm / 3 d)	Discharge (m³ / d)
15/01/2008	0.6	2.0	0.00	16/04/2008	0.0	0.4	0.00
16/01/2008	0.8	2.8	0.00	17/04/2008	0.0	0.2	0.00
17/01/2008	1.0	2.5	0.00	18/04/2008	0.0	0.0	0.00
18/01/2008	1.8	3.6	0.00	19/04/2008	0.0	0.0	0.00
19/01/2008	3.0	5.7	0.00	20/04/2008	0.0	0.0	0.00
20/01/2008	3.3	8.0	0.00	21/04/2008	0.0	0.0	0.00
21/01/2008	0.9	7.1	0.00	22/04/2008	0.0	0.0	0.00
22/01/2008	1.2	5.4	0.00	23/04/2008	0.0	0.0	0.00
23/01/2008	0.5	2.7	0.00	24/04/2008	0.0	0.0	0.00
24/01/2008	7.6	9.3	0.00	25/04/2008	0.0	0.0	0.00
25/01/2008	23.6	31.7	0.00	26/04/2008	0.0	0.0	0.00
26/01/2008	10.9	42.1	4.73	27/04/2008	0.0	0.0	0.00
27/01/2008	0.7	35.2	3.97	28/04/2008	0.0	0.0	0.00
28/01/2008	0.8	12.4	0.00	29/04/2008	0.0	0.0	0.00
29/01/2008	0.3	1.8	0.00	30/04/2008	0.0	0.0	0.00
30/01/2008	0.3	1.4	0.00				
31/01/2008	0.2	0.8	0.00				

7. A water balance modelling approach to optimising the use of water resources in ephemeral sand rivers[*]

7.1. Abstract

Alluvial aquifers present a possibility for conjunctive use with surface reservoirs for the storage of water in ephemeral sand rivers, such as the Mzingwane River in the Limpopo Basin, Zimbabwe. The Lower Mzingwane valley is a semi-arid region with high water stress, where livelihoods have revolved around the large rivers for thousands of years. However, current water allocation favours the commercial user: Of the 2,600 ha irrigated in the 5,960 km^2 (596,000 ha) region, only 410 ha are for smallholder farmers. A water balance approach was used to model the surface water resources and groundwater resources to determine the potential for expanding irrigation and to explore water allocation options. Using a combination of field and laboratory investigations, remote sensing and existing data, the Lower Mzingwane valley was modelled successfully using the spreadsheet-based model WAFLEX, with a new module incorporated to compute the water balance of alluvial aquifer blocks. Results showed that the lower Mzingwane alluvial aquifers can store 38×10^6 m^3 of water, most of that storage being beyond the reach of evaporation. Current water usage can be more than tripled: the catchment could supply water for currently-planned irrigation schemes (an additional 1,250 ha), and the further irrigation of two strips of land along each bank of the Mzingwane river (an extra 3,630 ha) – without construction of any new reservoirs. The system of irrigating strips of land along each bank of the Mzingwane river would be decentralised, farmer or family owned and operated and the benefits would have the potential to reach a much larger proportion of the population than is currently served. However, there could be substantial downstream impact, with around nearly one third of inflows not being released to the Limpopo River. The approach developed in this paper can be applied to evaluate the potential of alluvial aquifers, which are widespread in many parts of semi-arid Africa, for providing distributed access to shallow groundwater in an efficient way. This can enhance local livelihoods and regional food security.

7.2. Introduction

7.2.1. The Potential of Alluvial Aquifers in Ephemeral Rivers

In semi-arid areas without perennial rivers, smallholder farmers often have very little access to blue water for irrigation (Love et al., 2006a) and local storage of water is becoming an essential adaptation, with a growing consensus that small-scale water

---

[*] Based on: Love, D.; van der Zaag, P.; Uhlenbrook, S.; Owen, R. 2010c. A water balance modelling approach to optimising the use of water resources in ephemeral sand rivers. *River Research and Applications*, **26**, 908-925. DOI: 10.1002/rra.1408

supply technologies are the most cost-effective (Lasage *et al.*, 2007; Van der Zaag and Gupta, 2008). Runoff harvesting can make up some of the deficit, and has been used since ancient times in the higher rainfall areas of central and eastern Zimbabwe (Mwenge-Kahinda *et al.*, 2007; Love and Walsh, 2009), but this is less practical in areas with low rainfall and flat topography, such as the Lower Mzingwane Subcatchment in southern Zimbabwe. Here, the occurrence of sand rivers presents a distributed, environmentally-friendly option for cumulative storage for water supply.

Sand rivers offer an alternative to conventional surface water reservoirs for storage (Quillis *et al.*, 2009). The beds of sand rivers form alluvial aquifers: unconfined, horizontally semi-continuous layers of sand, silt and clay, forming the river channel, banks and, in places, flood plains. Because of the aquifers' shallow depth and intimate relationship with the river, flow in alluvial aquifers is essentially an extension of surface flow (Mansell and Hussey, 2005). Recharge of these aquifers is continuous for perennial rivers, but the more common case is annual recharge from an ephemeral river (Barker and Molle, 2004). Thus, in ephemeral systems, an alluvial aquifer provides the sole source of blue water during the season when the river does not flow. Evaporation only occurs in the near-surface sand: the extinction depth has been determined at 0.6 to 0.9 m below the surface (Nord, 1985; Mansell and Hussey, 2005; Aerts *et al.*, 2007). Storage of water in an alluvial aquifer has thus the potential of greater efficiency than storage in a surface dam (lower evaporation losses), as well as avoiding the environmental impacts associated with dams, such as inundation and flow regime changes (King *et al.*, 2003). The operations of dams can also have a negative impact on alluvial aquifers: continued channel incision, caused by the changed flow regime associated with the dam (Allan and Castillo, 2007) can eventually drain aquifers downstream of the dam (Bornette and Heiler, 1994) and sediment deposition downstream of the dam may be greatly reduced (Kondolf and Swanson, 1993; Shields *et al.*, 2000), partly due to declining channel migration (Ward and Stanford, 1995).

The dimensions of alluvial aquifers tend to be enhanced at natural geological boundaries due to differential erosion rates on the different lithologies either side of the boundary. This results in an increase in the storage potential through increasing respectively either the depth or width of the aquifer over the softer lithology downstream or upstream of the boundary (Owen and Dahlin, 2005; Cobbing *et al.*, 2008). Similarly, the construction of sand dams or gabion weirs can enhance the depth of the aquifer when constructed above surface (e.g. Aerts *et al.*, 2007) and minimise downstream groundwater flow when constructed subsurface (e.g. De Hamer *et al.*, 2008). Recharge of the aquifers by managed releases from an upstream dam, also increases the year-round availability of alluvial groundwater (Moyce *et al.*, 2006).

As has been discussed in section 6.2 above, alluvial aquifers present an important opportunity for increasing smallholder irrigation and thereby improving food production.

7.2.2. General Conceptual Model of an Alluvial Aquifer

The behaviour of an alluvial aquifer can be illustrated through a simple conceptual water balance model. The water balance can be expressed as:

$$Q_{in,s} + Q_{in,g} = E_s + Q_{abs} + G + Q_{out,s} + Q_{out,g} + \frac{dS}{dt} \qquad (7.1)$$

Where $Q_{in,s}$ and $Q_{in,g}$ are surface water and groundwater inflows respectively, E_s is nett evaporation from saturated sand (i.e. after consideration of infiltration of all precipitation into the aquifer), Q_{abs} is abstraction, G is seepage from the base of the alluvial aquifer (deep percolation), $Q_{out,s}$ and $Q_{out,g}$ are surface water and groundwater outflows respectively and $\Delta S/\Delta t$ is change per time step in storage in the alluvial aquifer. All units are volume fluxes per time step, such as $m^3 d^{-1}$ or alternatively depth fluxes, such as mm a^{-1}. In some geological environments, especially in younger crystalline rock terrains, the bedrock is impermeable and seepage from the aquifer is negligible: stream channels tend to erode and remove weathered regolith, leaving fresh, less permeable bedrock at the base of the sand bed. However, in older, more deeply-weathered terrains, seepage can be a substantial flux (De Hamer *et al.*, 2008); it is best estimated by monitoring water level declines during periods when $Q_{in,s}$ and Q_{abs} are zero. Groundwater outflow only occurs where there is another alluvial aquifer block immediately downstream of the one being studied and surface water outflow only occurs when the aquifer is fully saturated.

A small alluvial channel aquifer (area = 3 ha, depth = 4 m, hydraulic conductivity = 26.8 m d^{-1}, specific yield = 0.14, porosity = 43 %) is simulated in Figure 7.1. In this simple case, the aquifer commences at the upstream end of the model (thus excluding groundwater inflow), a geological barrier prevents groundwater outflow and impermeable bedrock prevents seepage. Evaporation losses were calculated using the surface area of the alluvial aquifer and reference evapotranspiration using CROPWAT for Windows 4.2 (Clarke, 1998). Two natural surface water inflows per year are simulated and it is assumed that direct recharge from rainfall is negligible. Simulation A shows the aquifer under natural conditions: the water table drops due to evaporation, until it goes below 0.90 m at which point it stabilises. When there is a surface flow event, e.g. 18 December 2007, the aquifer is fully saturated, after which evaporation acts upon it again. Simulation B shows the effect of demand for a small irrigation scheme (area = 4 ha). The aquifer empties 90 days after the start of the simulation and 90 days after the January flow events (the two recharge events of January 2008 in Figure 7.1). Simulation C shows the effect of conjunctive use of surface water and groundwater: recharging the aquifer by managed releases from an upstream dam: four releases are sufficient to ensure the irrigation scheme can be supplied year-round. There is thus a trade-off between the greater water supply available from managed releases against the evaporation losses in the dam upstream and the environmental impact of that dam. The major fluxes are summarised in Table 7.1.

Table 7.1. Major fluxes in the hypothetical small alluvial aquifer.

Scenario	A	B	C
Change in storage (cm a^{-1})	21	-345	-137
Abstractions (cm a^{-1})	0	776	1 321
Recharge from river (cm a^{-1})	94	482	1 259
Evaporation losses (cm a^{-1})	74	51	75
Porosity	43%	43%	43%

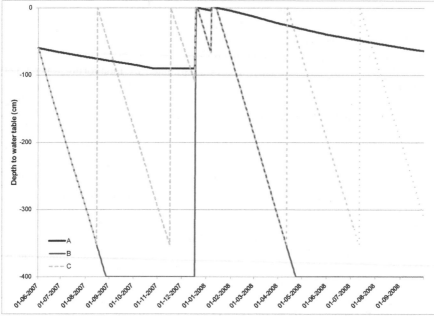

Figure 7.1. General behaviour of a small alluvial aquifer above impermeable bedrock (area = 3 ha, depth = 4 m), simulated using equation (1). A: the aquifer in its natural, unutilised state; B: the aquifer now supplying a 4 ha irrigation scheme; C: the aquifer supplying a 4 ha irrigation scheme and recharged by managed releases from a hypothetical upstream dam. Simulated using data from the Mushawe alluvial aquifer (Love *et al.*, *submitted*) – see the inset of Figure 7.2 for location.

7.2.3.   The Lower Mzingwane Subcatchment

The Lower Mzingwane Subcatchment (Figure 7.2) is a 5,955 km^2 semi-arid region with high water stress: potential evaporation (1,800 mm a^{-1}) is four times the mean rainfall (between 360 and 465 mm a^{-1}) (Mupangwa *et al.*, 2006; Love *et al.*, in press). Rainfall is seasonal, controlled by the Inter Tropical Convergence Zone and falling between October and April (Makarau and Jury, 1997). Rainfall occurs over a limited period of time, and in a small number of events (De Groen and Savenije, 2006). Because of these stresses, life and livelihoods in southern Zimbabwe have revolved around the larger rivers since ancient times and given birth to cultures such as Mapungubwe (Manyanga, 2006). To improve temporal availability of water from these ephemeral rivers, several large dams have been constructed, mainly designed to supply irrigation water (Table 7.2). Despite this, there are currently only 414 ha of smallholder irrigation schemes in the Lower Mzingwane Subcatchment. These schemes grow mainly maize and wheat, with some horticultural production - generally three crops per year. Some are supplied by surface water, others by alluvial groundwater. Some planned smallholder schemes have not yet been constructed, e.g. the Mtetengwe Irrigation Scheme.

Table 7.2. Major dams in the study area subcatchment. The capacities shown are based upon the most recent siltation surveys and are in some cases below the design capacity of the dam. Source: MCC (2009). For locations, see Figure 7.2.

Dam	River	Dam Capacity (10^6 m^3)	Constructed
Silalabuhwa Dam	Insiza	23.0	1966
Doddieburn Dam	Silonga	9.8	1970s
Tongwe Dam	Tongwe	3.3	1970s
Makado	Umchabezi	12.3	1980s
Siwaze	Siwaze	2.2	1980s
Zhovhe	Mzingwane	133.0	1995
Oakley Block	Mzingwane	40.7	Not constructed

Although the Lower Mzingwane Subcatchment thus has limited surface water availability, it does have the most extensive alluvial aquifers of all the tributaries in the Limpopo Basin (Görgens and Boroto, 1997). The alluvial aquifers form ribbon shapes covering over 99 km in length and 40 km^2 in areal extent along the Mzingwane river. Infiltration rates are fairly uniform spatially, due to the physical homogeneity of alluvium. Recharge of the alluvial aquifers is generally excellent and is derived principally from river flow (Owen, 1991). Adjacent to the riverbed is a narrow (less than 1 km wide) non-continuous flood plain of alluvial sediments, which are older than the current riverbed, based on field relations. This belt also hosts alluvial groundwater, with an estimated storage capacity of 22x10^6 m^3 (Moyce et al., 2006), but which has been found to be saline. They are characterised by high levels of sodium and chloride, which is an ambient condition, related to the geology of the aquifers and threatens irrigated agriculture with equipment or crop failure. (Love et al., 2006b). Little is known about the flood plain aquifers, although farmers report saline intrusion from them into the riverbed aquifer during periods of heavy pumping, or during periods when there is no recharge to the riverbed aquifer.

The riverbed alluvial aquifers represent an important and largely-untapped water resource, with current exploitation limited to 1,640 ha of commercial irrigation between Zhovhe Dam and Beitbridge. These commercial farms abstract water from boreholes and well-points in the river bed and on the banks for the irrigation of export crops, mainly citrus, and are resupplied by release and spillage of water from Zhovhe Dam and annual ephemeral river flow, which recharge the aquifer (Love et al., 2008a). Zhovhe Dam also releases water for Beitbridge town, which abstracts the water from the Limpopo River, into which the Mzingwane flows. The dam was built with the intention of supplying surface water to smallholder irrigation schemes (not yet developed) and to the town of Beitbridge, as well as recharging the alluvial aquifer used by the commercial farmers.

There are 25 small dams in the study area, ranging in capacity from 70,000 to 5,000,000 m³ and together representing 16 % of the surface water storage by volume in the study area.

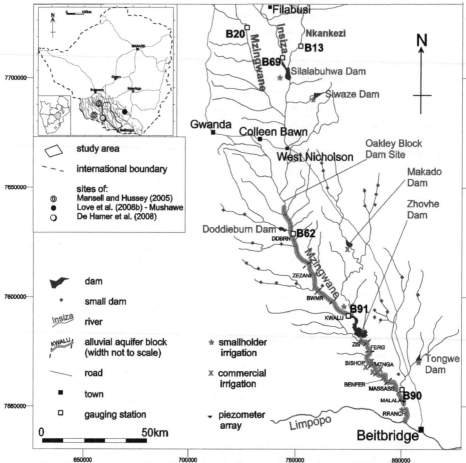

Figure 7.2. Map of the Lower Mzingwane valley and lower Insiza valley, showing locations of dams, alluvial aquifer blocks, gauging stations and towns. Inset: location (shaded) within Zimbabwe and southern Africa. Base map after Moyce *et al.* (2006). The aquifer blocks are defined by impermeable boundaries such as rapids or virtual boundaries – the boundaries of farms.

7.2.4. Objectives

A water balance approach was used to model the surface water resources and alluvial aquifers of the Mzingwane river system downstream of gauges B20, B13 and B69, in order to (i) develop a modelling method that could provide information for development planning from limited data, (ii) assess whether demand from planned schemes can be met without additional storage infrastructure, (iii) evaluate whether further expansion of smallholder irrigation, using riverbed alluvial aquifers, is possible without additional storage infrastructure, and (iv) determine the benefits of construction of an additional large dam at Oakley Block. The Insiza Subcatchment was also considered as a possible source for managed releases from Silalabuhwa Dam.

A problem encountered in this study was scarce data availability and the fact that most tributaries of the lower Mzingwane River are ungauged – a not uncommon situation in many semi-arid areas of the world. To work with the available data, certain assumptions were necessary. Where theywere made, the assumption giving a more conservative estimate of water resource availability was selected, to ensure that the water available for allocation in the model was not over-simulated.

7.3.   Methods

7.3.1.   Data Sources

Rainfall and discharge data were obtained for the longest possible period where data was available and of sufficient quality from all required stations (gauging stations B13, B20, B69 and rainfall stations at Beitbridge, Filabusi and West Nicholson). This allowed for use of the following data periods: 1966-1977, 1981-1983, and 1987-2003. Examination of annual rainfall data (Figure 7.3) shows that most of the highest and lowest rainfall years since 1950 are represented. Although gauging stations do exist on the Mzingwane downstream of B20 (in the DDBRN and KWALU blocks on Figure 7.2), data availability is sketchy and does not overlap well with data availability from B13, B20 and B69. Data is also compromised by siltation of the gauge towers. Given the difficulties of estimating the characteristics of ephemeral river systems (Hughes, 2005), it is unwise to make use of data sets of poor quality or with limited timespan.

For ungauged tributary catchments, the discharge of the largest gauged but undeveloped catchment was used, namely the Nkankezi River at gauging station B13. The discharge recorded at B13 was regionalised by applying coefficients for catchment area and climatic region:

$$Q_x = abQ_{B13} \qquad\qquad (7.2)$$

$$a = \frac{A_x}{A_{B13}} \qquad\qquad (7.3)$$

Where Q_x is discharge for ungauged tributary x (million cubic metres per ten days: $10^6 m^3 (10d)^{-1}$), Q_{B13} is discharge recorded at gauging station B13 ($10^6 m^3 (10d)^{-1}$), A_x is the catchment area of ungauged tributary x (km^2), A_{B13} is the catchment area of gauging station B13 (km^2) and b is a coefficient (-) that is used to take into account the fact that the ungauged catchments of the study area lie in lower rainfall areas than B13. Linear regression of unit discharge from gauging station B90 with that from B13 (see Figure 7.2

for locations and proximity to climate stations) provided a value for *b* of 0.51 (correlation coefficient = 0.64). Gauging station B90 is in the lowest rainfall part of the study area (near Beitbridge) and is downstream of some small dams (see Figure 7.1), so this comparison should give a conservative estimate of discharge in an ungauged tributary.

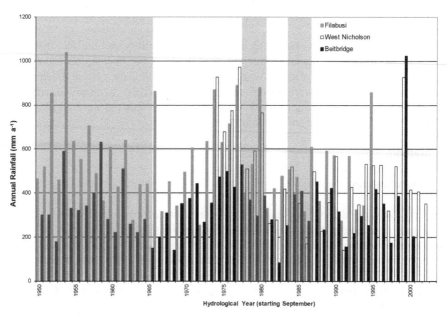

Figure 7.3. Annual rainfall data from the study area, after Love *et al.* (2010a). The data periods used in this study are shown by a white background. The rainfall stations are located in towns shown in Figure 7.2.

The model domain does not incorporate the Mzingwane River upstream of gauging station B20, or the Insiza River upstream of gauging station B69 for two reasons. First, there are no alluvial aquifers on these river reaches. Second, the water resources of these subcatchments are fully committed, and in some tributaries overcommitted (Kabel, 1984). The upper Mzingwane and Insiza are fully developed for the supply of water to Bulawayo, Zimbabwe's second largest city (MCC, 2009; Kileshye-Onema *et al.*, 2006). Given the strategic importance of Bulawayo's water supply (Gumbo, 2004), it is unrealistic to consider changes to water allocation from the upper Mzingwane or upper Insiza for downstream use. Instead the approach followed was to use actual historic discharges at B20 and B69, which are downstream of the abstractions made for Bulawayo. Note that the data periods selected, as discussed above, continue into the period that post-dates the most recent dam constructions.

Details on data used are given in Tables 3 and 4. Daily data were converted to a dekad (10 days) time step, by adding the daily flux values. For data only available at a monthly time step, it was assumed that the parameter did not vary within the given month and the monthly value was thus divided into the appropriate number of dekads. Annual data was handled in a similar fashion and thus the values for such data were constant throughout the time series.

Cattle population density in the area is sparse (e.g. World Resource Institute, 2000; Thornton *et al.*, 2002). Neither the estimated livestock water requirements for Zimbabwe (e.g. De Hamer *et al.*, 2008) nor the more generous ones for South Africa (e.g. KwaZulu-Natal Department of Agriculture and Environmental Affairs, 2008) suggest a significant annual livestock water demand: For the entire study area an annual livestock water demand was computed from these figures as $0.27 \times 10^6 \text{m}^3 \text{a}^{-1}$, spread across the twenty-five small dams shown in Figure 7.2. This demand was less than 14% of the evaporation losses from these dams (see below) and was therefore ignored. Water demand was also ignored for the dams on two game farms where the only consumptive use was domestic and each farm had a population of under 100 (field observations, this study).

Table 7.3. Data used in the model.

Type of data	Source
Rainfall	Meteorological Services Department daily data from Beitbridge, Filabusi and West Nicholson.
Evaporation	Hargreaves formula (Allen *et al.*, 1998) using Meteorological Services Department daily temperature and radiation data from West Nicholson.
Discharge	(i) Zimbabwe National Water Authority hydrology database (daily). (ii) For ungauged tributary catchments, discharge of the undeveloped Nkankezi River (B13) was used, adjusted per equations (1) and (2).
Demand: irrigation	Varies with crop mixture and calendar month: (i) Interview with farmers and District authorities to establish crop patterns and extent. (ii) Crop water requirements modelled using CROPWAT for Windows 4.2 (Clarke, 1998) at a monthly time step
Demand: urban	MCC (2009) – annual data, assumed to be identical in each dekad.
Dam locations and surface areas	(i) Zimbabwe National Water Authority permit database. (ii) Planning documents (MCC, 2009; SNV, 2001). (iii) Additional dams not recorded in the permit database of planning documents were captured from Landsat scene path170 row075 dated 3 December 2000 and the wet season surface area digitised.
Dam capacities and volume/ area relationships	(i) Zhovhe Dam rating table, Zimbabwe National Water Authority. (ii) Maximum capacities of some dams available from permit database and planning documents as above. (iii) Relationships for all dams except Zhovhe were estimated using method *b* of Sawunyama *et al.* (2006) and capacities, where known. This method gives the largest estimate of area and therefore of evaporation loss, which is conservative for water resource availability.

Hydrogeological parameters used in this study are shown in Table 7.4. The experimentally derived values for specific yield are well below the average value used by Owen and Dahlin (2005) in their study in the BWMR block and that of Cobbing *et al.* (2008) used in their study of the Limpopo River; thus this study gives a more conservative estimate of aquifer capacity. There was incomplete information for some aquifer blocks, so in those cases the value from the adjacent block (given the same bedrock lithology) for the missing parameter was used. Given the large spatial scale of the model, possible heterogeneities in hydrogeological parameters within a given block or cross-section were ignored.

Table 7.4. Hydrogeological parameters for the Lower Mzingwane alluvial aquifers. See Figure 7.2 for locations. Storage capacity was computed from the other parameters.

Block	Length (km)	Surface area (km²)	Depth (m)	Specific yield (-)	Hydraulic conductivity (m d⁻¹)	Storage capacity (10⁶ m³)
DDBRN	3.21 [a]	7.80 [a]	4 [g]	0.107 [e]	41.7 [h, i]	3.35
ZEZANI	13.50 [a]	4.35 [a]	4 [d]	0.077 [i]	41.7 [i]	1.33
BWMR	9.90 [a]	3.97 [a]	5 [f]	0.077 [i]	41.7 [i]	1.52
KWALU	12.61 [a]	5.33 [a]	13 [k]	0.077 [i]	41.7 [h, i]	5.30
ZIS	3.22 [a]	0.96 [a]	13 [j]	0.107 [e]	31.6 [e]	1.33
FERG	10.68 [a]	4.20 [a]	12 [c]	0.107 [d]	31.6 [d]	5.87
BISHOP	4.05 [a]	2.33 [a]	20 [c]	0.090 [d]	46.2 [d]	4.19
MZNGA	5.97 [a]	2.69 [a]	20 [c]	0.090 [e]	46.2 [e]	4.84
BENFER	11.32 [a]	3.73 [a]	20 [c]	0.090 [d]	46.2 [d]	6.69
MASSASS	9.92 [a]	2.14 [a]	10 [b]	0.043 [d]	44.0 [d, h]	0.91
MALALA	5.87 [a]	0.98 [a]	15 [b]	0.043 [e]	44.0 [e, h]	0.63
RRANCH	9.08 [a]	2.65 [a]	15 [d]	0.043 [d]	44.0 [d]	1.69
Total	**99.34**	**41.13**	---	---	---	**37.65**

Data sources: [a] analysis of 3,4,5 band false colour composite of Landsat scene path170 row075 dated 3 December 2000, this study; [b] Field resistivity measurements, this study; [c] Interviews with farmers operating deep boreholes, this study; [d] Assumed similar to block immediately upstream, given similar bedrock geology (interpretation of Landsat scene specified above and field observations, this study); [e] Laboratory tests, this study; [f] Minimum value from Owen and Dahlin (2005); [g] Field observations, this study; [h] Block terminates at rapid or other impermeable barrier – field observations, this study; [i] No data available: used average of other blocks; [j] Calculated from Zhovhe Dam construction drawings; [k] Assumed similar to block immediately downstream, given similar bedrock geology (interpretation of Landsat scene specified above and field observations, this study).

Observations of water levels in piezometers in the alluvial aquifer at sites in ZIS, MZNGA and MALALA blocks showed that, over the course of the dry season, with the exception of one piezometer which was located close to a wellpoint, the water level never dropped more than 0.9 m below the river bed – the evaporation extinction depth. The absence of any drop in water levels once the water table dropped below this depth suggests that it is reasonable to treat seepage losses as negligible. This finding was also made by Owen and Dahlin (2005) at Bwaemura (BWMR block on Figure 7.2) and Mansell and Hussey (2005) at two sites in the adjacent Shashe river basin (locations shown on inset of Figure 7.2).

7.3.2. The WAFLEX Model and Model Development

WAFLEX is a simple and user-friendly model and runs in a spreadsheet (Savenije, 1995). WAFLEX has been applied extensively in southern Africa for the modelling of water allocation: between Swaziland, South Africa and Mozambique on the transboundary Incomati River Basin (Nkomo and Van der Zaag, 2004; Juízo and Líden, 2008), environmental flow requirements of the Odzi River in Zimbabwe (Symphorian et al., 2003) and water quality and mass balance of the Kafue River in Zambia (Mutale, 1994).

In WAFLEX each river reach, demand node or reservoir is a cell. In supply mode, each cell contains a simple formula to add water flowing into it from adjacent cells, and to subtract any demand connected to that cell. The demand mode is a "mirror-image" to the supply mode, and computes the required flow in an upstream direction, starting with the demand located furthest downstream. In this study, WAFLEX was run in Microsoft Excel. Macros use the formulae in the networks and input data to model each time step and generate simulated discharge at specific river reaches, and time series of abstractions and shortages. For each subcatchment, a network was schematized, containing all major tributaries, dams and demand nodes (see Table 7.5 and Figure 7.4).

In WAFLEX, a dam consists of three cells; an inflow cell, a storage cell and a release cell. The release cell acts as an inflow point to the downstream reach. The storage and release of the dam is determined in a macro subroutine, taking into account the flood rule curve and the dead storage curve. The storage can never exceed the former and the latter may never be crossed as a result of a release. Net evaporation from a reservoir is calculated at each time step from rainfall, pan evaporation data and the reservoir surface area, which is determined from the reservoir's area-volume curve. All the dams shown on Figure 7.2 were included in the model through separate macros and simulated individually.

In this study, WAFLEX was adapted to incorporate alluvial aquifers into the water balance, through a simple water balance of the alluvial aquifer block and overlying river reach. The macro treats an alluvial aquifer block as a storage unit, similar to a reservoir, with inflow, outflow and storage cells. The storage cell is a unit with a fluctuating volume of water, as calculated using equation (7.1) and the outflow is calculated by using Darcy's Law:

$$Q_{out,g} = -KA + \frac{dh}{dL} \tag{7.4}$$

Where $Q_{out,g}$ is groundwater outflow, K is hydraulic conductivity (metres per ten days: $10^6 m^3 (10d)^{-1}$), A is cross-sectional area (m²) and dh/dL is the gradient of the water table. The sources for the parameters of equations (1) and (4) are shown in Table 7.6.

Table 7.5. Demand nodes in the Lower Mzingwane Subcatchment

	Water user	Purpose	Total annual demand ($10^6\,m^3\,a^{-1}$)
1	Sukwi irrigation scheme	Agriculture	0.41
2	Zezani business centre	Urban	0.10
3	Bwaemura irrigation scheme*	Agriculture	2.06
4	Kwalu irrigation scheme	Agriculture	2.57
5	Watson	Agriculture	1.08
6	Zhovhe irrigation scheme	Agriculture	0.04
7	Park	Agriculture	7.20
8	Cawood	Agriculture	1.51
9	Cunliffe	Agriculture	1.38
10	Smith	Agriculture	0.69
11	Ferguson	Agriculture	1.56
12	Silalabuhwa irrigation scheme	Agriculture	4.53
13	Silalabuhwa growth point	Urban	0.001
14	Colleen Bawn	Mining	2.95
15	West Nicholson business centre	Urban	0.08
16	River Ranch irrigation scheme	Agriculture	0.10
17	Avoca business centre	Urban	0.01
18	Siwaze irrigation scheme	Agriculture	0.21
19	Tongwe irrigation scheme	Agriculture	1.03
20	Malala irrigation scheme*	Agriculture	1.89
21	Mtetengwe irrigation scheme*	Agriculture	8.75
22	Beitbridge [†]	Urban	2.89

* Planned or projected demands
[†] Beitbridge has an off-river storage capacity of $6.95\ 10^6\,m^3$

When the water stored in an aquifer block falls below 30 % of capacity, the aquifer module requests water supply from an upstream dam in the demand mode. This is to counteract problems that may be experienced at low water table levels, such as pump intake positioning or un-evenness of bedrock (Nord, 1985).

A time step of one dekad (ten days) was selected. WAFLEX has no routing module, so a daily time step was not realistic and one dekad is thus the finest temporal resolution possible.

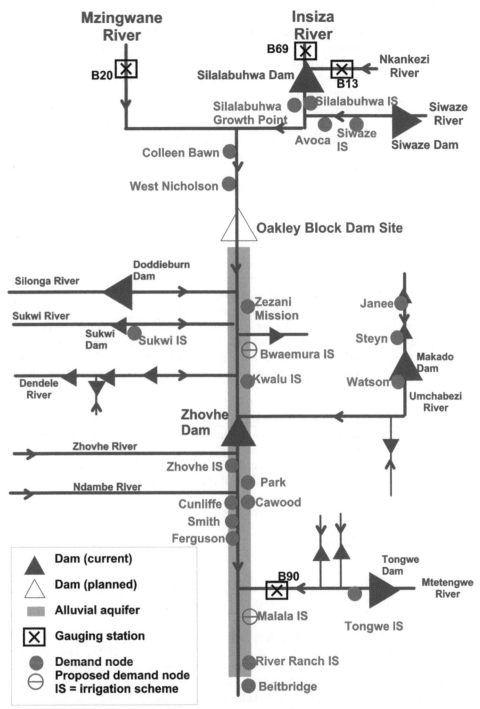

Figure 7.4. Schematisation of the lower Mzingwane River.

Table 7.6. Parameters for the alluvial groundwater module, incorporated into WAFLEX.

Parameter	Source
Inflow (10^6 m^3 $(10d)^{-1}$)	Total outflow of the upstream cell. If the upstream cell is an alluvial aquifer block, this includes groundwater flow and surface water flow
Evaporation (10^6 m^3 $(10d)^{-1}$)	Derived from regional evaporation values (Table 7.2), applicable only to that portion of the aquifer less than 0.90 m below the surface and saturated for a given time step. Evaporation extinction depth is based upon evaporation loss experiments in sand: water levels make a curve asymptotically approaching 0 evaporation as the depth to water approaches 0.90 m (Nord, 1985). This is deeper than that used by Mansell and Hussey (2005) or Aerts *et al.* (2007), thus giving a generous estimate of evaporation and a more conservative estimate of water resource availability.
Abstractions (10^6 m^3 $(10d)^{-1}$)	The default value for each time step was derived from demand (Table 7.2): abstractions for a given time step will be less than the default value if the computed storage, after subtraction of evaporation, is less than the demand.
Seepage	Treated as negligible, as discussed above.
Groundwater flow (10^6 m^3 $(10d)^{-1}$)	A default value for each block was calculated using equation (4). Groundwater flow will be less than the default value if the computed storage for a given time step, after subtraction of the preceding parameters, is less than the default value. There is no groundwater flow out of RRANCH, MASSASS and MALALA blocks, which were found to terminate downstream against surface exposure of bedrock.
Surface flow (10^6 m^3 $(10d)^{-1}$)	If after subtraction of the preceding parameters, the computed storage exceeds the capacity of the aquifer, the balance is expressed as surface flow. Users of surface water can only access this volume, which will be zero unless the aquifer is fully saturated.
Cross-sectional area	From depth (see Table 7.4) and field surveys of the channel width
Hydraulic conductivity (m $(10d)^{-1}$)	See Table 7.4.
Gradient (-)	Assumed to be the same as the gradient of the riverbed, which was calculated from 1:50,000 topographic mapping.

It is not suggested that this modelling approach is comparable with the precise numerical approach of groundwater models such as MODFLOW (Harbaugh, 2005). The latter are ideal for small-scale modelling of sites with comprehensive field data available (Hughes *et al.*, 2010). However, for regional studies of this nature, a simpler approach is needed – and in data-scarce areas a simple water balance approach may be all that is possible.

Validation of the model was complicated by limited availability of independent data in general, and the absence of any hydrological gauging station on the main Mzingwane River downstream of Zhovhe Dam. Independent discharge data were available for limited time periods at two gauging stations upstream of Zhovhe Dam: B62 (1981-1988) and B91 (1985-1991). The model was therefore validated separately against the aforementioned two discharge series and against groundwater levels in aquifer block MALALA, recorded during the course of this study (2008 – for location of the piezometer array see Figure 7.2).

A simple sensitivity analysis was carried out for the best parameterisation for the eight parameters that were varied during the calibration. The 10 % elasticity index (equation 4.12) was used.

7.4.   Results and discussion

7.4.1.   Model Validation and Sensitivity

The model performed well when validated against independent discharge data (Table 7.7). Simulated and observed groundwater levels were well-correlated, and the low Nash-Sutcliffe coefficient could be due to the fact that the observed groundwater levels are from one site in the MALALA aquifer block and the simulated levels are for the whole block – local variation cannot be reproduced by the model.

The sensitivity analysis (Table 7.8) showed that the model is not very sensitive to the main hydrogeological parameters, nor to the storage levels of the dams at the start of a model run.

Table 7.7. Results of model validation

Data source	Discharge, B62	Discharge, B91	Groundwater levels, MALALA
R Correlation Coefficient	0.73	0.82	0.95
R^2 Determination Coefficient	0.53	0.67	0.91
C_{NS} Nash Sutcliffe Coefficient	0.52	0.64	0.36
n Number of time steps	131	125	23

Table 7.8. Local sensitivities of model outputs to model parameters, calculated using equation (4.12).

Model output	Elasticity index (e_{10}), first figure given is for 10 % increase in the model parameter, second figure is for 10 % decrease							
	Specific Yield		Hydraulic Conductivity		Evaporation extinction depth		Proportion of dam capacity full at start	
Outflow	-0.07	0.07	0.00	0.00	0.00	0.00	0.35	-0.35
Total evaporation losses	-0.01	0.01	0.00	0.00	0.02	-0.02	0.05	-0.07
Shortage amount	0.00	0.00	0.00	0.00	0.00	0.00	0.00	0.00

7.4.2. Scenario Modelling

Scenario development can be an effective tool in water resources planning and management, although where there is considerable uncertainty in future developments (Dong *et al.*, in press). In this study, six discrete scenarios were modelled for specific future developments using 30 years of data: the three time periods shown in Figure 7.3 were combined, see Table 7.9 and Figures 5 and 6. All demands can be met in the baseline scenario (Scenario 0), and also in Scenario 1, where the planned irrigation schemes annotated in Table 7.4 are also supplied, bringing the total area irrigated to 3,847 ha (additional irrigation is assumed to be for maize, crop water requirements modelled using CROPWAT for Windows 4.2 (Clarke, 1998)). No significant shortage occurs, including during all of the drier years.

In Scenarios 2 to 5, a narrow belt of smallholder irrigation is envisaged on each bank of the Mzingwane River, for the full length of the river where the alluvial aquifers are developed, i.e. from block DDBRN to the confluence with the Limpopo River, 99 km long. In Scenario two, 3,633 ha of land can be irrigated in strips 190 m wide along each river bank, in addition to the irrigated area considered in Scenario 1. Demand is met except on three occasions, in each case the second year of a two-year drought. Even in that case, the cumulative shortage at Beitbridge (the user furthest downstream) is much lower than the town's off-river storage capacity. When drought years follow a year with normal inflows (or are followed by such a year), demand is fully met.

Scenario 3 considers the same arrangement, but without the existence of Zhovhe Dam. It can be seen that in years with normal inflows, all current and planned demands can be met throughout, as well as 287 ha additional irrigation, on strips along each river bank. Scenario 4 (construction of Oakley Block Dam without the existence of Zhovhe Dam) allows this figure to rise to 1,338 ha (70 m wide strips). During drought years there is insufficient water to supply the river bank irrigation and downstream releases from Silalabuhwa Dam empty the dam, sacrificing water supply for Silalabuhwa Irrigation Scheme and resulting in shortage for users of the upstream alluvial aquifers as well.

Table 7.9. Res

Description of

Area irrigate

		In
River flows		Ou
		Da at
		Cc
Storage		Da
		Ac
Evaporation losses		Da
		Ac
		To
		Ot
		Ot Cc e

Descript	Consumptive use	Shortage

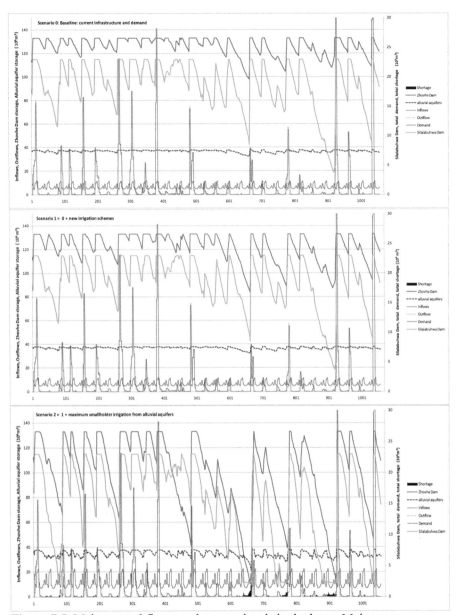

Figure 7.5. Major annual fluxes and storage levels in the lower Mzingwane Catchment under three of the six scenarios as modelled in WAFLEX.

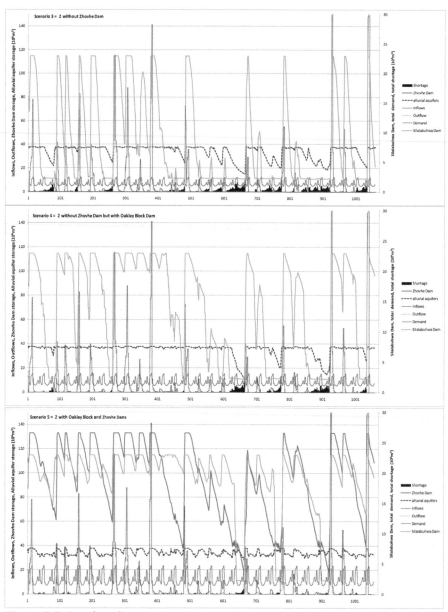

Figure 7.5. (continued)

In scenario 5, construction of Oakley Block Dam on the current scenario (i.e. with Zhovhe Dam existing) was found to allow for 4,015 ha of irrigation along the river (on strips 210 m wide along each river bank) with small shortages recorded during the driest years. Cumulative shortage for aquifer block RRANCH during the longest shortage period is lower than Beitbridge town's off-river storage capacity in all scenarios. Evaporation losses from the dams and aquifers are between 3 and 4 % of inflows, except for Scenario 3 which excludes the largest dams (Zhovhe and Oakley Block). Scenario 2 is the most efficient in that evaporation losses are equivalent to only around 12 % of the volume of productive use of water. Downstream impact in terms of flow reduction is currently around 10 %. However, for scenarios 2 and 5 around 30 % of inflows are consumptively used (abstracted and evaporated). The number of days the Mzingwane River flows at its mouth drops from the current 103 per year to 82 in scenario 2.

Annual evaporation losses modelled for Zhovhe Dam were 3.0×10^6 m^3 a^{-1} and 1.3×10^6 m^3 a^{-1} for Silalabuhwa Dam. For the small dams losses ranged from 0.01 to 0.3 $\times 10^6$ m^3 a^{-1} each and gave an average annual total for the twenty-five dams of 2.04×10^6 m^3 a^{-1} (6 % of their total storage). This is substantially more than the maximum proportion estimated to be consumed from these dams by livestock $(0.27 \times 10^6$ m^3 a^{-1} , that is 0.8 % of their total storage). This moderate level of evaporation loss from small dams is consistent with results from the West African savannah (Volta Basin: Andreini *et al.*, 2009).

7.4.3. Behaviour of the Alluvial Aquifer

It could be shown for the study area that in unutilised or minimally utilised aquifer blocks, groundwater flow generally exceeded evaporation by a factor of at least 10. In the more heavily used aquifer blocks, evaporation was minimised: for BISHOP, the most heavily used block in scenario 0, the average abstraction was 0.20×10^6 m^3 (10d)$^{-1}$ compared to 0.27×10^6 m^3 (10d)$^{-1}$ in groundwater outflow. The groundwater outflow is thus a major component of the water balance and plays an important role in recharging the downstream aquifer blocks and allowing for abstractions from these blocks during the dry season – a finding also made by Mansell and Hussey (2005) and Owen and Dahlin (2005). The groundwater outflow rate of an aquifer block is a function of the hydraulic conductivity, which is a function of the physical properties of the aquifer material (river sediment) and the water table gradient. If the composition of new sediment changes, as may be happening immediately downstream of Zhovhe Dam due to retention of sediments by the dam (see supplementary material), the groundwater outflow rate will also change.

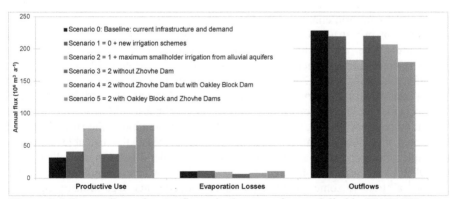

Figure 7.6. Comparison of water fluxes in the scenarios modelled in WAFLEX, normalised to catchment area. Annual inflows were 256 10^6 m^3 a^{-1} for every scenario.

7.4.4. Advantages and Disadvantages of the Water Balance Model Approach

Regional groundwater models, while not having the higher levels of data richness and lower levels of uncertainty associated with detailed site investigations, do have an important role to play in water resources planning (Palma and Bentley, 2007).

Simplified modelling of aquifer storage and fluxes at regional level allows for the identification of development opportunities (Cobbing *et al.*, 2008). However, development investment must be preceded by more detailed site investigations. This study was limited by lack of some hydrogeological data from most aquifer blocks (see Table 7.3). Field investigations in the future could reveal opportunities not identified in this study – but could also invalidate certain assumptions made. Nevertheless, it is suggested that the conservative approach followed in making assumptions and selecting data sources is sufficient to produce reasonable minimum values on water resource availability and irrigation potential.

7.5.   Conclusions and Recommendations

7.5.1.   *WAFLEX model and Alluvial Groundwater*

The poor data availability that is typical of this environment and the assumptions limited the scope of this study. However, the WAFLEX model, as adapted to include alluvial aquifers, proved a suitable and flexible model that can provide useful information for planning purposes from limited data – and has also produced a comparable trend in findings on alluvial aquifer behaviour to published field studies.

Given the choice of assumptions and data sources giving a more conservative estimate of water resource availability, it is suggested that this study's findings represent robust values for the minimum areas that can be irrigated under the different development scenarios. It is recommended that a suitable rainfall-runoff model should be regionalised for this area, to replace the crude regionalisation of equations (2) and (3) and thereby improve the robustness of the inflow data used in the model.

7.5.2.   *Maximising Smallholder Irrigation (Scenario 2)*

Despite the semi-arid climate of the Lower Mzingwane catchment, the current water management infrastructure is grossly under-utilised. This paper has demonstrated that current water use (including 2,597 ha of irrigation) can be tripled with existing infrastructure: the catchment could supply water for currently-planned irrigation schemes (an additional 1,250 ha), and the further irrigation of two strips of land along each bank of the Mzingwane river (an extra 3,633 ha). The additional capacity can provide for 4,350 households at 0.29 ha per household from the planned irrigation schemes and 12,700 households from irrigation along the river. It should be noted that poor irrigation efficiency would decrease the irrigable area. For example, the crop water requirements for smallholder irrigation modelled using CROPWAT for Windows 4.2 were 10,366 m^3 ha^{-1} a^{-1}, which is similar to the 10,000 m^3 ha^{-1} a^{-1} used in some government planning documents (e.g. Ministry of Local Government Rural and Urban Development, 1996), but higher figures are sometimes used in planning (e.g. 16,000 m^3 ha^{-1} a^{-1} for Mwenezi-Dinhe; MCC, 2009). Using a higher figure could decrease the prospective irrigable area. However, since part of the transmission losses from irrigation carried out along the riverbank are likely to be returned to the aquifer, the net loss from the system would be less.

Demand is more easily met through the use of Silalabuhwa and Doddieburn Dams to resupply the alluvial aquifers, whilst still supplying their existing users. However, there could be substantial downstream impact, with around nearly one third of inflows not being released to the Limpopo River. The system of irrigating strips of land along each bank of the Mzingwane river, provided suitable soils are available, would be decentralised, farmer or family owned and operated and the benefits would have the potential to reach a much larger proportion of the population than is currently served. This functional and operational decentralisation of smallholder irrigation should lead to benefits in terms of management and water use efficiency (Jaspers, 2003) as well as cost savings through the use of cheaper technology at the smaller scale (Love et al., 2006a). It has been seen in India and Bangladesh that direct control of water supply by individual farmers gives them better flexibility and reliability than the centralised approach, and led to rapid development of conjunctive surface water – groundwater use in irrigation (Faurés et al., 2007) and improved access to water for women (Van Koppen and Mahmud, 1996). It has been shown elsewhere in Africa that farmers can switch rapidly to irrigation from groundwater, and invest in appropriate irrigation technology where it makes economic sense (Shah et al., 2007). This could include small motorised pumps, or even manual pumps such as the rower pump where head differences are not too large. However, state support is still required for security of access to land and water, monitoring and agricultural extension services (Faurés et al., 2007).

This water allocation scenario would also maximise productive use of water compared to evaporation losses from dam and aquifer storage. Such conjunctive use of surface water and groundwater is more efficient than the traditional approach where alluvial aquifer recharge is considered as "transmission losses" (e.g. MCC, 2009). The frequent releases from Zhovhe Dam would minimise the intrusion into the alluvial aquifer blocks BISHOP, MZNGA and BENFER of slightly saline water from the flood plain aquifers, which is a problem for irrigation water supply especially during dry years (Moyce et al., 2006).

Evaporation losses modelled for small livestock dams were found to be far more significant than the small amount of water consumed by the livestock for which the dams were constructed.

7.5.3. Development Implications of Other Scenarios

Current demand could have been met without the construction of Zhovhe Dam. The currently-planned irrigation schemes could also be supplied (1,250 ha) and a similar amount along the banks of the river (1,256 ha) – without the environmental impact from the construction and operation of Zhovhe Dam.

A substantial change in the flow regime of the Mzingwane River occurs downstream of Zhovhe Dam (see supplementary material), which could be the cause of observed changes in channel morphology, where the active riverbed has declined in width and portions of the river channel on either side have been abandoned and appear to be being colonised by vegetation such as Acacia xanthophloea. The ongoing non-inundation and abandonments of portions of the river channel and the apparent loss in aquifer material indicate that the extent of the alluvial aquifer is likely to decrease. The possible change in grain size could result in a decline in specific yield. Such changes would negatively impact the water storage and water supply potential of the Lower Mzingwane alluvial aquifer, and thus the current and potential water users. However, consideration of the environmental impact of this dam study is constrained by the lack of an environmental baseline prior to construction. (Love et *al.*, 2008a). At the same time, it should be noted that the dam has had a positive social impact, including employment and the development of a small fishing industry, in addition to the obvious water supply benefits.

Utilisation of the alluvial aquifers alone would have been a cheaper and more environmentally-friendly option than the construction of Zhovhe Dam – its evaporation losses (2.4×10^6 m^3 a^{-1}) are equivalent to the annual demand of Beitbridge town. Use of the current off-river storage capacity at Beitbridge is sufficient to guarantee the town's supply under all scenarios. However, in drought years users of Silalabuhwa Dam experience shortage as that dam empties through downstream releases.

Construction of Oakley Block Dam instead of Zhovhe Dam could have supplied the current and currently-planned demand as well as 1,338 ha along the banks of the river. The considerably smaller size of Oakley Block Dam compared to Zhovhe Dam suggests that its environmental impact and financial cost would have been lower – although no environmental impact assessment or cost/benefit analysis has been carried out for Oakley Block Dam.

Construction of Oakley Block Dam in addition to Zhovhe Dam could supply the current and currently-planned demand as well as an additional 4,015 ha along the banks of the river. It should be noted that this is only about 400 ha more than could be supplied by Zhovhe Dam alone. Given the small gain in irrigated area, the construction of Oakley Block Dam is not attractive, considering the possible environmental impact of the dam (although likely less than that of Zhovhe) and its evaporation losses of 0.8×10^6 m^3 a^{-1}. The remote (and uninhabited) location of the site mitigates against the development of the type of fishing industry that exists at Zhovhe Dam.

7.5.4. *Future Research and International Implications*

Alluvial aquifers are widespread in many African countries, such as Botswana, Namibia, Nigeria and South Africa. This study has shown that such aquifers may represent an alternative to large dam construction that is less costly financially and environmentally. A thorough investigation of possible water supply from alluvial aquifers should be part of the scoping exercise when considering the possible construction of a dam. Low cost manual abstraction systems can be applicable due to the small head difference between the alluvial bed and the fields (Mansell and Hussey, 2005) and the resource availability will often mirror population concentrations along the larger rivers. This is particularly important in semi-arid southern Africa within the context of climate change and possible declines in the frequency and total amount of rainfall (Shongwe *et al.*, 2009; Love *et al.*, 2010b). This is an important area for future research.

The modified WAFLEX modelling approach followed in this study allows for a sound calculation of water availability in the river-aquifer system, which is important for development planning. The approach can be applied to evaluate the potential of alluvial aquifers in semi-arid Africa for providing distributed access to shallow groundwater in an efficient way. This can enhance local livelihoods and regional food security.

7.6. Supplementary Material: Impact of the Zhovhe Dam on the lower Mzingwane River channel

Extracted from: Love, D., Love, F., van der Zaag, P., Uhlenbrook, S. and Owen, R.J.S. 2008. Impact of the Zhovhe Dam on the lower Mzingwane River channel. In: Humphreys, E., Bayot, R.S., van Brakel, M., Gichuki, F., Svendsen, M., Wester, P., Huber-Lee, A., Cook, S., Douthwaite, B., Hoanh, C.T., Johnson, N., Nguyen-Khoa, S., Vidal, A. and MacIntyre, I. (eds.). *Fighting Poverty Through Sustainable Water Use: Proceedings of the CGIAR Challenge Program on Water and Food 2nd International Forum on Water and Food, Addis Ababa, Ethiopia, November 10 – 14 2008*, **IV**. The CGIAR Challenge Program on Water and Food, Colombo, pp46-51. [ISBN 9789299005323]

The lower Mzingwane River was dammed in 1995 by the large (133×10^6 m^3) Zhovhe Dam. Managed releases from the dam supply commercial agro-business and Beitbridge town, downstream. A substantial change in the flow regime of the Mzingwane River occurs downstream of Zhovhe Dam, with the capture of all flows early in the rainy season, most low flows, many larger flows and the reduction in magnitude of some floods. This change in flow regime could be the cause of observed changes in the channel morphology, where the active riverbed has declined in width and portions of the river channel on either side have been abandoned. These abandoned parts of the river channel appear to be being colonised by vegetation, which represents competition for water with the established riparian vegetation as well as with the water users. The ongoing non-inundation and abandonments of portions of the river channel and the apparent loss in aquifer material indicate that the extent of the alluvial aquifer is likely to decrease. The change in grain size from a coarse sand to a fine gravel (if substantiated, further sampling is required) could result in a decline in specific yield. Such changes would negatively impact the water storage and water supply potential of the Lower Mzingwane alluvial aquifer, and thus the current and potential water users. It is widely acknowledged that indigenous vegetation has evolved to prevailing water balance and soil moisture conditions (Farmer *et al.*, 2003), the riparian ecosystem, notably an important stand of acacia woodland, is also likely to suffer.

7.6.1. Methods

The flow regime upstream and downstream of Zhovhe Dam was compared for the period since construction of Zhovhe Dam (1995 – 2008). This comparison was made since there are no data for discharge downstream of the dam site prior to construction. The downstream discharge was obtained from records kept at the Dam but there was inadequate upstream data: the principle upstream gauging station (B91) had unreliable and incomplete data for the period since the construction of Zhovhe Dam and there were no records available for flow in the Mchabezi River that is principal tributary flowing into the dam. The upstream flow was therefore reconstructed using a water balance of the dam:

$$Q_{inflow} + P_{ZhDm} = E_{ZhDm} + G + Q_{spill} + Q_{rel} + \Delta S/\Delta t \qquad (7.4)$$

Where Q_{inflow} = river inflow to the dam: flow at gauging station B91 plus flow at the mouth of the Umchabezi River (total; m^3 d^{-1}), P_{ZhDm} = rainfall on Zhovhe Dam (m^3 d^{-1}; derived from rainfall records, dam level records and the dam surface area – volume curve), E_{ZhDm} = evaporation from Zhovhe Dam (m^3 d^{-1}; derived from pan evaporation, dam level records and the dam surface area – volume curve), P_{ZhDm} =

rainfall on Zhovhe Dam (m^3 d^{-1}; derived from rainfall records, dam level records and the dam surface area – volume curve), G = seepage (m^3 d^{-1}; average estimated from days with no inflows, outflows or releases; found to be consistent with the generally low permeability of basalts of this formation in southern Africa, except where the basalt layers are thin or heavily fractured (Bell and Haskins, 1997; Larsen et al., 2002), Q_{spill} = discharge over the spillway (m^3 d^{-1}; calculated from dam levels records kept at the dam, using the standard equation for a broadcrested weir), Q_{rel} = discharge through the managed release valves (m^3 d^{-1}; records kept at dam), $\Delta S/\Delta t$ = change in storage per time step (m^3 d^{-1}; derived from dam level records).

To study the geomorphology and vegetation, six sites were selected based on regional geological mapping, confirmed by ground truthing. The sites covered three distinct types of bedrock: granitoid, sandstone and basalt. One site was selected upstream and one downstream for each of the three lithologies (see Figure 7.7). A cross-section was surveyed and the vegetation along the cross-section identified at each site. An historic comparison of vegetation, using satellite images from five years before construction and five years after was attempted, using a false colour composite using bands 3, 4 and 5 and a NDVI composite from Landsat scene path170 row075 from May 1990 and June 2000.

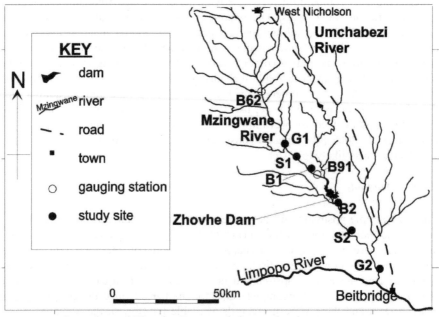

Figure 7.7. Location of Zhovhe Dam and the study sites.

At site S2, 30 km downstream of the dam, a grain size analysis of the channel material was available from before the construction of Zhovhe Dam (Ekström et al., 1996). A grain size analysis was carried out in 2008 with material sampled from the same locality.

7.6.2. *Results and discussion*

Flow regime

In most years upstream discharge increases from late November / early December but is retained by Zhovhe Dam in the earliest part of the rainy season (October/November), see Figure 7.8. This is typical of large dams in the northern Limpopo Basin (Kileshye-Onema *et al.* 2006). Spilling occurs leading to downstream discharge in late December / early January. Sometimes the late rainy season flows (April/May) are also retained, e.g. April 2008. All upstream flows are retained by the dam during drier years. For example, the only downstream flows in the 2006-2007 climatic season were planned releases for irrigation – this was a drought year with a moderate El Niño recorded during that season (Logan *et al.* 2008).

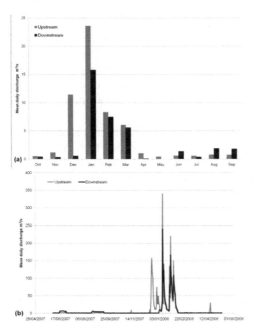

Figure 7.8. Flow regime of the lower Mzingwane River upstream and downstream of Zhovhe Dam: (a) Mean monthly values since construction of Zhovhe Dam in 1995, (b) discharge recorded for the 2007-2008 season.

There are no downstream low flows during the rainy season: the only downstream flows are the larger floods when the dam spills. The loss of low flows has important ecological significance, for example, low flows are thought to sustain herbaceous riparian vegetation (King *et al.* 2003). Such changes in flow regime can also be associated with loss of habitat and biodiversity (Brown and King 2003).

The dam releases only a limited number of large floods from upstream. The magnitude of these floods is often decreased, including for some of the extreme events such as Cyclone Japhet, which occurred during a season with low rainfall and some of the inflowing water was retained by the dam. This contrasts with the operation of some other large dams in Zimbabwe, such as Insiza and Rusape, where an increase in size or number of large floods is observed downstream (Kileshye-Onema *et al.* 2006; Love *et al.* 2006c).

There are small managed releases are made during June/July and August/September for irrigation and other water users; this is during the dry season when there are no natural (upstream) flows.

Geomorphology and vegetation

On both sandstone and basalt, the downstream cross-sections show greater channelization (Figure 7.9). At the downstream sandstone site Mazunga Drift (site S2) part of the sand bed on each bank has now been abandoned by the river and the active channel width is reduced from nearly 200 m at Bwaemura School (site S1) to 150 m. The abandoned parts of the channel have been colonised by shrubs. At the downstream basalt site Zhovhe Irrigation Scheme (site B2) the active channel is confined to an incised bed of under 30 m along the left bank, with a sloping flood channel of some 150 m, compared to an active bed width of 250 m upstream (Kwalu, site B1). The bulk of the sand bed is now apparently being colonised by trees and shrubs. The narrowing of the active (bare) river channel and the colonisation of abandoned parts of the channel can be clearly seen at sites S2 and B2

The difference is less apparent on the granitoids, although the channel is narrower at G2 than G1. It is likely that the observed channelization is due to (i) the regular system of managed releases, and (ii) the decline in larger floods, and decline in magnitude of floods. This results in only partial inundation of the river channel. The analysis of satellite imagery proved inconclusive.

Sedimentology

Comparison of grain size analyses (Figure 7.10) carried out at Mazunga Drift (site S2) immediately before (1995) and thirteen years after (2008) the construction of Zhovhe Dam show a substantial decline in the finer sediments and increase in the larger particle fraction. A X^2 test confirmed that the sample from site S2 in 2008 is statistically distinct from that of site S2 in 1995 at 95 % probability. However, the 2008 samples from sites S2 and B2 were confirmed as statistically similar. Material from site B2 (2008) is presented for comparison. Note that the composition of material sampled from site S2 in 2008 is much closer in composition to that sampled from site B2 on the same date, than to that sampled from site S2 at the earlier date (1995).

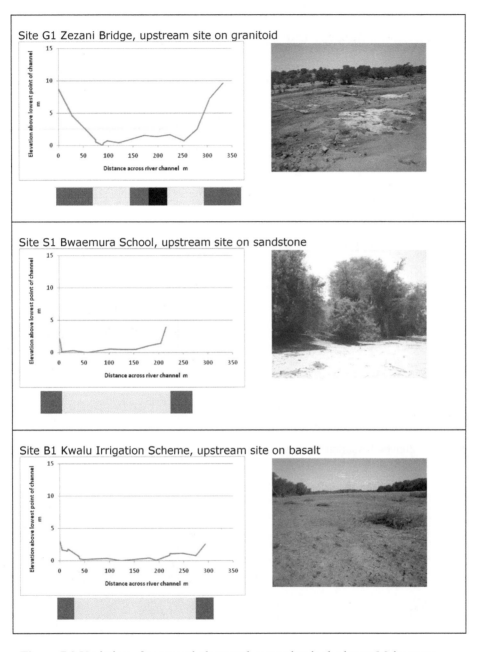

Figure 7.9. Variation of geomorphology and vegetation in the lower Mzingwane River channel for sites with similar geology, upstream and downstream of Zhovhe Dam. For each site: (left) surveyed cross-section at study site, with vegetation along cross-section below. (right) photo from study site. Key to vegetation: green = trees/shrubs, yellow = clear riverbed (no vegetation), light green = sandbank with trees/shrubs, black = bare rock.

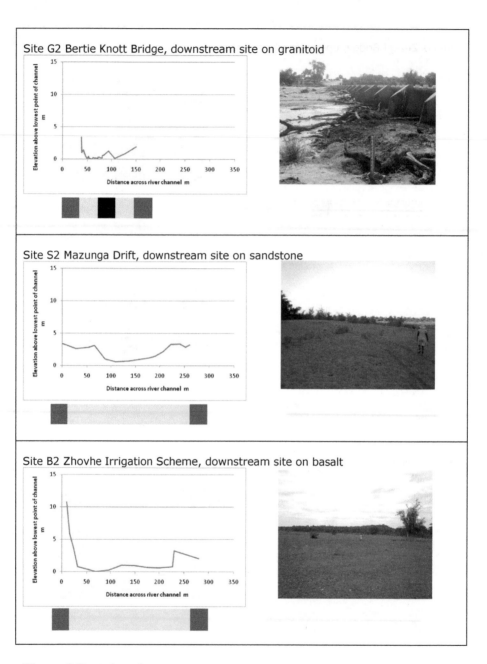

Figure 7.9 continued

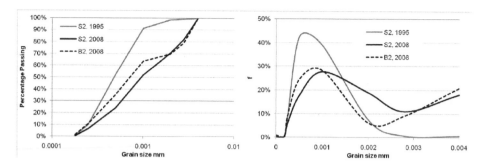

Figure 7.10. Grain size analysis of channel bed (alluvial aquifer) material at Mazunga Drift (site S2), immediately before (1995) and thirteen years after (2008) the construction of Zhovhe Dam. 1995 data after Ekström *et al.* (1996).

This change at site S2 from 1995 to 2008 is from a mainly coarse sand sediment to a fine pebble / very coarse sand sediment, and a substantial change in grain size distribution. There are two possible explanations for this observation: The first explanation is that the 2008 sediment represents substantively different material. It is to be expected that selective removal of fines from the Mzingwane River sediment-load takes places due to the relatively clean, sediment-free lower energy water that spills over the dam. Such water would be expected to have been derived from a different provenance: neither the spillway nor the managed release system at Zhovhe Dam allow for significant downstream release of sediments, thus more recent (post-1995) sediments in the riverbed will have been carried down into the Mzingwane River by tributaries. There are two right-bank tributaries that join the Mzingwane River downstream of Zhovhe Dam (Figure 7.7). Thus although erosion and removal of the existing (and pre-dam) downstream sediments continues, their replacement is from a different source and appears to favour the larger coarse sand / fine gravel particles.

The alternative explanation is that the 2008 sediment represents material from a different depositionary environment: The sample might have been collected from a slightly different point than the 1995 sample, or the alignment of the river channel could have changed in the intervening period. In either case, the 2008 sample could represent a different depositionary environment, thus different sedimentary facies. If this explanation is true, then no useful interpretations can be made from the observed differences. The two explanations are not mutually exclusive.

8. Water Resource Constraints and Opportunities to Increasing Dryland Food Security

8.1. Abstract

Southern Africa faces the twin challenge of declining water resource availability – from lower rainfall and runoff – and rising water demand – mainly from irrigated agriculture which must expand to meet food, fuel and fodder needs in particular from the rapidly growing urban areas. Governments, development agents, NGOs and individual farmers are all developing strategies and interventions to improve food security and rural livelihoods. Most of these strategies and interventions have an effect on the water cycle, through increasing the demand for surface and groundwater, through changing land and agricultural water use or both. The challenge and its response thus operate in the context of stressed water resource availability. Nowhere is this clearer than in the semi-arid lands, such as the Limpopo Basin.

In this study, a coupled hydrology and water resources model is used to evaluate land/water scenarios and livelihood intervention strategies in the Insiza Catchment, Limpopo Basin. The model used integrates a conceptual rainfall-runoff model (HBVx; Love et al. 2010a) with a water resources assessment model (WAFLEX; Savenije 1995). HBVx is adapted from HBV-light (Seibert 2002) with an interception storage and flux pre-processor and WAFLEX with a simple groundwater balance module. The model evaluates climate change and intervention responses such as changing crop choice and irrigation method, conservation agriculture, new irrigated lands and crop-livestock integration through their effects on soil/water dynamics, runoff generation, aquifer recharge and water allocation.

The model's performance is adequate, although change simulated in many scenarios is below soil parameter sensitivity. The model demonstrates the more than proportional response of discharge to a decline in rainfall – but also the benefits of change in crop type and irrigation method. Substantial livelihood gains can be realised by switching from rainfed maize production to cattle fodder crops and small grains, with minimal effects on the simulated catchment water balance. Either change could potentially benefit around 16,000 households. It was estimated that irrigation area can be increased by up to 19%, benefitting around 1,800 households, through use of alluvial aquifers, multiple use of existing small reservoirs and construction of new ones. However, these options have significant downstream impact, with catchment outflows decreasing by up to 7%. Important efficiency gains can be made from conjunctive use of reservoirs (where upstream dams release water when the storage in the dam downstream falls below a threshold), and from drip irrigation.

8.2. Introduction

Southern Africa faces the twin challenge of declining water resource availability and rising water demand. Water resource availability is declining as rainfall and runoff decrease (Love et al. 2010b) and is likely to do so further as the impact of climate change on water resource availability is felt (Desanker and Magadza, 2001; Milly *et al.*, 2005; 2008; Andersson *et al.*, 2011), especially in Botswana and Zimbabwe (Engelbrecht *et al.*, 2011). Other changes such as a delay in the onset and early cessation of the rainy season, and an increase in the severity of droughts can also be expected (Shongwe *et al.*, 2009).

Water demand is increasing, mainly from irrigated agriculture, which must expand to meet food, fuel and fodder needs, but also due to the water supply requirements of rapidly-growing urban areas (e.g. Van der Zaag and Gupta, 2008; Ncube *et al.*, 2010; Komakech *et al.*, 2012) and initiatives to improve access to water for the rural poor (Dlamini, 2008). Demand for irrigation water will rise as climate change reduces dryland crop production (Stige *et al.*, 2006) – but also as agriculture is a major priority for economic growth in Sub-Saharan Africa (Commission for Africa, 2005). Expansion of irrigated agriculture and development of groundwater and alternative water sources are priorities for southern Africa in the Southern African Development Community's climate change strategy (SADC, 2011).

Governments, development agents, NGOs and individual farmers are all developing strategies and interventions to improve food security and rural livelihoods. Most of these strategies and interventions have an effect on the water cycle, through increasing the demand for water resources (surface water and groundwater), through changing land use or both (Falkenmark and Rockström, 2004; Love *et al.*, 2006a; Moyo *et al.*, 2006; Mupangwa *et al.*, 2006; Hanjra and Gichuki, 2008; Vidal *et al.*, 2009). The challenge and its response thus operate in the context of stressed water resource availability. Nowhere is this clearer than in the semi-arid lands, such as the Limpopo Basin.

The objectives of the study are: (i) to determine the water resources implications of certain development and livelihood intervention scenarios in terms of changes in catchment inflow, outflow, evaporative losses and productive use of water, and (ii) to determine the possible impact of climate change on the water resources of the study area in terms of changes in catchment inflow, outflow, evaporative losses and productive use of water.

8.3. Methods

8.3.1. Study Area

This study was carried out in the Insiza Catchment, a major left-bank tributary of the Mzingwane River. The Mzingwane Catchment is thus a secondary catchment (of the Limpopo Basin) and the Insiza Catchment a tertiary catchment. The 3,401 km^2 catchment (Figure 8.1) is underlain by gneissic and granitic rocks, giving rise to moderately shallow, coarse grained kaolinitic sands, and moderately shallow clays and loams (Bangira and Manyevere, 2009; Chinoda *et al.*, 2009). Land use is mainly a mixture of croplands, pastureland and woodland. The woodland is open and evolves from *Combretum–acarcia–terminalia* in the north to *Mopane* in the south and is widely used for livestock pasturing. Cropping includes commercial farming (largely resettled) in the north, often under irrigation, and smallholder farming (mostly rainfed) in the south (Kileshye-Onema *et al.*, 2006). Free-range cattle ranching is common throughout the catchment. The proportion of land under cropping is increasing and the area under good natural woodland is declining (Kileshye-Onema and Van Rooyen, 2007).

The Insiza River and its tributaries have been dammed at numerous locations, to provide water for urban areas, mines and irrigation (Table 8.1). The flow regime of the Insiza River is modified by these dams, particularly in regard to low flows (Kileshye-Onema *et al.*, 2006). There are also at least 30 smaller dams (capacity 0.25 to 2 x 10^6 m^3, number derived from ZINWA records and examination of Landsat images), which provide water for livestock, smallholder irrigation, nutritional gardens, fishing and domestic use (Sawunyama *et al.*, 2005; 2006; Senzanje *et al.*, 2011). The lower reaches of the Insiza River host alluvial aquifers, whose potential for water supply has only been marginally developed (Moyce *et al.*, 2006).

Table 8.1. Major dams in the study area subcatchment, with a capacity of 2 x 10^6 m^3 or more. The capacities shown are based upon the most recent siltation surveys and are in some cases below the design capacity of the dam. Source: Mzingwane Catchment Council (2009). For locations, see Figure 8.1.

Dam	River	Dam Capacity (10^6 m^3)	Constructed
Silalabuhwa Dam	Insiza	23.4	1966
Upper Insiza (Fort Rixon) Dam	Insiza	8.9	1967
Insiza (Mayfair) Dam	Insiza	173.4	1973
Siwaze Dam	Siwaze	2.2	1980s

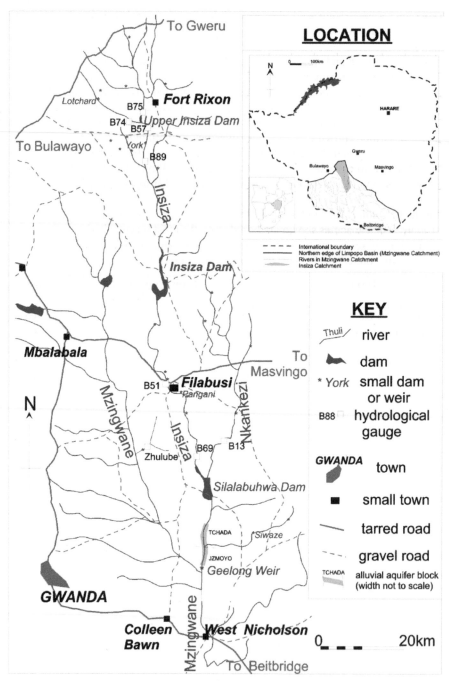

Figure 8.1. The Insiza Catchment, showing locations and gauges used in this study. HBVx, the modified version of HBV-light (Seibert 2002) used in the coupled model was developed by Love *et al.* (2010a) in the Zhulube catchment, in the centre of the study area. Inset: location in Zimbabwe and southern Africa.

8.3.2. Data sources

The major sources of data for this study are given in Table 8.2. The data selected for modelling and validation cover the period January 1980 to December 2000. This has been selected on the basis of data availability and covers a wide range in climatic variation, as can be seen in Figure 8.2). It should be noted that this period is also after the major change points in water resources availability determined in Chapter 3, and is therefore likely to be more representative of current water resource availability conditions.

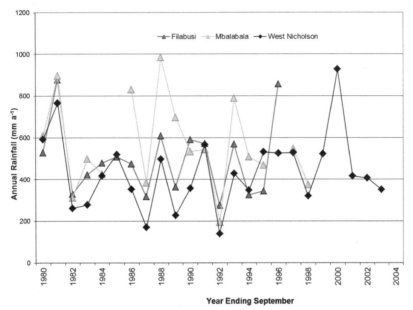

Figure 8.2. Annual rainfall of selected climate stations in the study area, 1980/81 to 2003/04. For station locations see Figure 8.1.

The quality of input data is of high importance since this influences both model performance and the parameter sets to be regionalised. There is a minimum quantity of input data required for model parameterisation. Where input data is limited by missing values or large uncertainties, interpretations are seriously compromised (Seibert and Beven, 2009). The high spatial and temporal variability in rainfall, and runoff in the study area exacerbates this problem. The time series were visually inspected, along with supporting materials such as the station files. The following exclusions were made in order to ensure that the data used did not have years where rainfall or discharge data was missing or readings were unreliable: (i) where rainfall or discharge data was missing for two months or more. (ii) where rainfall or discharge data was missing for two weeks or more during the months of November to April (rainy season). (iii) where a note has been made in the station file that readings were unreliable (e.g. due to siltation, security).

For the purposes of rainfall-interception-evaporation-runoff modelling, the Insiza Catchment was divided into twenty-five sub-catchments (Table 8.3). Data used in the modelling of the alluvial aquifers is given in Table 8.4.

8.3.3. *Rainfall-interception-evaporation-runoff Modelling*

The HBV (Hydrologiska Byråns Vattenbalansmodell) family of models, whilst developed and applied initially in Sweden, have also been used in semi-arid and arid countries such as Australia, Iran and Zimbabwe (Líden, 2000; Líden and Harlin, 2000; Oudin et al., 2005; Masih et al., 2010). Love *et al.* (2010a) adapted HBV-light (Seibert, 2002) to incorporate an interception storage and flux pre-processor and to set up the model in Microsoft Excel (see sections 4.3.4 and 4.3.5).

The interception pre-processor calculates the potential evaporation at catchment level derived using mapped land use (Table 8.3) and crop coefficients (Table 8.5). Daily interception was then calculated following De Groen and Savenije (2006) with a threshold interception storage value of 5 mm. However, if the amount of rainfall intercepted was more than could be evaporated on that day, some moisture will remain in interception storage until the next day, thus decreasing the available volume of interception storage for that day (Love *et al.*, 2010a).

Table 8.5. Crop coefficients used for different land cover types, varying by season. For cultivated land, values for maize in East Africa (FAO, 2013b) were used.

	South African equivalent	Jan	Feb	Mar	Apr	May	Jun	Jul	Aug	Sep	Oct	Nov	Dec
Woodland: highveld	Woodland (indigenous tree/bush savanna) [a]	1.14	1.14	1.14	1.14	1.00	1.00	1.00	1.00	1.07	1.14	1.14	1.14
Woodland: mopane	Mopani veld [b]	0.74	0.74	0.71	0.57	0.54	0.47	0.43	0.50	0.57	0.64	0.69	0.71
Mixed grassland and woodland	Mixed bushveld [b]	1.00	1.00	0.93	0.86	0.71	0.64	0.57	0.64	0.79	0.93	0.93	1.00
Mixed grassland and woodland (degraded)	Veld in poor condition [c]	0.79	0.79	0.79	0.64	0.29	0.29	0.29	0.29	0.43	0.57	0.71	0.79
Wetland	Wetland grasses [c]	1.14	1.14	1.14	1.00	0.86	0.71	0.57	0.57	0.57	0.71	0.86	1.00
Rocky hills	Veld/rock 50 % to 100 % rock [c]	0.43	0.43	0.43	0.43	0.43	0.43	0.43	0.43	0.43	0.43	0.43	0.43

Source: [a] Jewitt, 1992; [b] Schulze and Hols, 1993; [c] Schulze et al., 1995. The original crop coefficients were derived for use with pan evaporation data (Schulze et al., 1995). These were converted for use with reference evaporation data by dividing the original crop coefficient with the pan coefficient (taken as 0.7).

The interception pre-processor operates as the first routine in HBVx, and is followed by the standard set up of HBV-Light: a soil moisture routine and a runoff generation routine with two reservoirs (Figure 4.3). The soil moisture routine is governed by the non-linear function β, which computes the amount of infiltration water that goes into runoff generation, and the maximum soil moisture storage FC (mm), which is similar to field capacity. The runoff generation routine comprises an upper, fast-reacting reservoir, and a lower, slow-reacting reservoir. Flow of water from these reservoirs into runoff is governed by the linear storage coefficient (K_0, K_1, K_2) and overflow from the upper storage (Q_0) based upon exceedance of the threshold storage volume UZL (mm). The parameters used in the model are explained in Table 5.5. The parameter values and a time step of one dekad (ten days) were selected from regionalisation of the HBVx model in the Mzingwane Catchment (Love et al., 2011). Such an approach is considered preferable to unguided automatic calibration which can give preposterous parameter values in semi-arid catchments (Líden, 2000).

The parameter set selected for the model (Table 5.5) was the one which performed best in regionalisation (dekadBC of Love et al., 2011), and was derived from catchments with similar physical characteristics to the Insiza Catchment.

8.3.4. Surface Water and Alluvial Groundwater Routing and Allocation Model

WAFLEX is a simple and user-friendly water resources model and runs in a spreadsheet (Savenije, 1995). Love et al. (2010c) adapted WAFLEX to incorporate alluvial aquifers into the water balance, through a simple water balance of the alluvial aquifer block and overlying river reach (see section 7.3.2 and equations 7.1 and 7.4). A time step of one dekad (ten days) was selected. WAFLEX has no routing module, so a daily time step was not realistic and one dekad is thus the finest temporal resolution possible and meaningful from a process point of view as river routing processes are negligible at this time step interval.

8.3.5. Coupled Model

With both HBVx and WAFLEX operating in Excel at a dekad time step, the models are coupled. The runoff output of HBVx becomes input into WAFLEX as the river flow in sub-catchments upstream of dams, and small tributary catchments (see Figure 8.3). The coupled model is referred to as HBVx On WAFLEX Spreadsheet Information Tool (HOWSIT). This tool is another illustration of the power spreadsheets have, allowing the modeller to tailor the model to the required purpose (Olsthoorn, 1985).

The interface (see Supplementary Material Figure 8.6) allows modification of interception storage, simulation of climate change or land use change in the pre-processor. Land management change can be simulated in the HBVx module interface and water use and reservoir sequencing in the WAFLEX module.

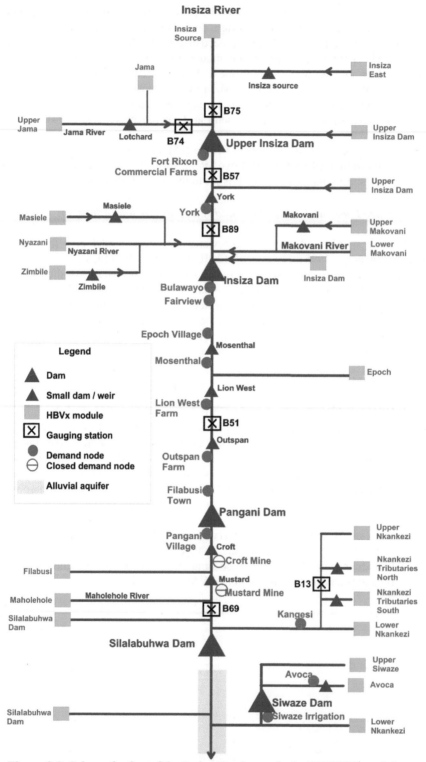

Figure 8.3. Schematisation of the Insiza Catchment in the HOWSIT model.

The model was validated against data from several gauging stations, using the Nash-Sutcliffe Coefficient (equation 4.10). A simple sensitivity analysis was carried out for the validated model. The 10 % elasticity index (equation 4.12) was used.

8.3.6. Scenario Modelling

The principal drivers of (undeveloped) semi-arid catchment behaviour are vegetation cover, the type and intensity of land use and climate (e.g. Farmer et al., 2003; Walker et al., 2006). Many of the interventions and opportunities for increasing dryland food production do so through changes in agricultural water management (e.g. Love et al., 2006a; Moyo et al., 2006; Mupangwa et al., 2006; 2012; 2013; Hanjra and Gichuki, 2008; Ncube et al., 2009; 2010; Vidal et al., 2010) and thus potentially change catchment behaviour. A series of scenarios were developed to explore these possible changes (Table 8.6), drawing extensively (but not exclusively) on research carried out in Challenge Program on Water and Food projects 17 (Ncube et al., 2010) and 46 (Andreini et al, 2009).

The major climatic changes expected for southern Zimbabwe as a result of global change have been well documented (e.g. Desanker and Magadza, 2001; Christensen et al., 2007) and comprise mainly a substantial decrease in rainfall and an increase in temperature. This is simulated in scenario 1 with a 10% decrease in rainfall and a 3 °C increase in temperature.

Land use is already changing, with a decrease in tree cover and forest, grazing land and wetlands, and an increase in area under dryland cultivation (Sibanda et al., 2011). Such changes can result in changes to infiltration rates (Ngwenya, 2006), increased overland flow and decreased water holding capacity (Mahe et al., 2005). These changes are simulated in scenario 3. These changes are often associated with land degradation, leading to soil erosion and deforestation (Milton et al., 2003). However, HOWSIT only simulates the effect of these pressures on discharge – not the major problems of, for instance, siltation and loss of biodiversity. The results of HOWSIT simulations of land use change are therefore likely to be minimum values, as the model does not simulate the compacting effect of over-grazing, which may also increase discharge (Adekalu et al., 2006), nor siltation, which decreases reservoir capacity, causing them to fill, spill and empty faster.

Many changes in dryland cultivation have been recommended. Switching from maize to the drought-resistant small grains such as sorghum and pearl millet, especially rapid-maturing varieties, is strongly recommended for increased food production through drought resistance (e.g. Cooper et al., 2008; Schlenker and Lobell, 2010; Thornton et al., 2010; Mupangwa et al., 2011), is primarily a change in crop choice, not in cultivation method. This is simulated in scenario 4A. Integrated crop-livestock systems where some of the cereal cultivation is replaced by fodder crops in order to increase household food and financial security (e.g. Hanjra and Gichuki, 2008; Homann and van Rooyen, 2008) also constitute a change in crop choice (scenario 5). Conservation agriculture to increase dryland cereal yields (e.g. Hanjra and Gichuki, 2008; Dhliwayo, 2006; Mupangwa et al., 2007; 2008; 2012; 2013) including in-field rainwater harvesting (e.g. Ncube et al., 2009; 2010) has potentially a more complex impact, with increases in soil water content (average of 10%) and decrease in surface runoff (average of 29%) expected (Mupangwa et al., 2008) – simulated in scenario 7.

Supplementary irrigation of rainfed agriculture, using river water, groundwater, or off-field rainwater harvesting has the potential to reduce the impact of droughts and dryspells on crop production (e.g. Magombeyi and Taigbenu, 2008; Ncube et al., 2010). However, the complexities of this intervention were not possible to simulate at the scale the rainfall-runoff model used in this study (HBVx) operates.

Interventions to improve the management of existing water resources systems can increase productivity of water resources at basin scale (Hanjra and Gichuki, 2008; Alam and Olsthoorn, 2011). In the study area, these could include conjunctive use of upstream and downstream reservoirs (Love et al., 2010c), where water is released from an upstream reservoir, when the water level in a downstream reservoir falls below a critical level, simulated in scenario 2. Multiple use (MUS) of existing reservoirs and water sources has major potential for livelihood upliftment and the expansion of irrigation (e.g. Van Koppen et al., 2006; Vidal et al., 2009; Senzanje et al., 2011). In this study, MUS simulated in scenario 8B by assigning an irrigation scheme to each dam which has none (the command area is based upon the relative sizes of Siwaze Irrigation Scheme and Siwaze Dam compared to the dam in question).

The use of alluvial aquifers for water supply and irrigated crop production has been demonstrated at other sites in the Limpopo Basin (e.g. De Hamer et al., 2008; Love et al., 2010c) and there is potential for this in the lower part of the Insiza Catchment, with over 300 ha of land irrigable along the short stretch of the lower Insiza River where alluvial aquifers are developed (simulated in scenario 6).

An important possible change in irrigated agriculture is the use of drip irrigation (e.g. Maisiri et al., 2005; Moyo et al., 2006; Ncube et al., 2010), either through conversion of existing irrigation schemes or development of new schemes. These possibilities are simulated in scenarios 9A and 9B.

A recent development in lowland irrigation in Zimbabwe is the conversion of irrigation schemes from smallholder maize to sugar cane for biofuel production, such as at Chisumbanje and Nuanetsi. This has important economic benefits at a national scale, but problematic livelihood impacts at the local scale (Mujere and Dombo, 2011; Scoones et al., 2012). The hydrological effects are simulated in scenario 4B.

Where there is scope to further develop water resources, investment in new irrigation schemes still presents a major benefit to rural communities, provided there is also sufficient investment in management, extension staff, linkage to markets and capacity building of the farmers (Faulkner et al.; 2008; Hanjra and Gichuki, 2008). The development of small reservoirs can have substantial livelihood benefits, provided they do not place an inequitable burden on downstream users (Andreini et al., 2009) and provided that the institutional capacity to manage the reservoirs is developed at the same time (Van der Zaag and Gupta, 2008). There are no sites registered in the outline plan (MCC, 2009) so the development of new reservoirs and associated water use is simulated with one new small dam with a storage capacity of 500,000 m^3 for each of the six sub-catchments of 200 km^2 or more in area in scenario 8A, and each with an irrigation scheme of 14 ha (the command area is based upon the relative sizes of Siwaze Irrigation Scheme and Siwaze Dam) in scenario 8B.

8.4. Results and Discussion

8.4.1. Model Validation and Sensitivity

The model was validated against discharge data for the period January 1980 to December 2000 (the model components have been previously calibrated: Love et al., 2010c; 2011). Validation gave mixed results, see Table 8.7, but simulated the flow well at most gauges on the Insiza River.

Table 8.7. Results of validation of HOWSIT model. For locations of gauges, see Figure 8.1.

Gauge	C_{NS}	R	R^2	Conclusion
B13	0.49	0.79	0.62	Good simulation of tributary
B51	-0.01	-0.05	0.00	Poor simulation of middle main river
B57	0.53	0.75	0.57	Good simulation of upper main river
B69	0.29	0.66	0.43	Moderate simulation of lower main river
B74	0.52	0.73	0.53	Good simulation of tributary
B75	-1.42	0.37	0.14	Poor simulation of tributary
B89	0.62	0.82	0.67	Good simulation of middle main river

Two of the main soil parameters: Beta and the maximum soil moisture storage (FC), gave up to 8% change in model output for 10% change in parameter, showing the importance of reliable parameter estimation and of land and soil management for hydrological processes. The model is not so sensitive to other parameters.

Table 8.8. Local sensitivities of model outputs to model parameters, calculated using Equation (4.12).

Elasticity index e_{10} (-), first figure given for each model output is for 10% increase in the model parameter, second figure is for 10% decrease								
Model / parameter output	Inflow		Outflow		Total evaporation losses		Shortage amount (all users)	
PERC [a]	-0.1	0.1	-0.1	0.1	-0.1	0.1	0.1	-0.1
UZL [a]	0.0	0.0	0.0	0.0	0.0	0.0	0.0	0.0
FC [a]	-0.3	0.4	-0.5	0.6	-0.3	0.3	0.7	-0.7
BETA [a]	-0.3	0.4	-0.5	0.6	-0.2	0.3	0.7	-0.8
Specific Yield	0.0	0.0	0.0	0.0	0.0	0.0	0.0	0.0
Dam start [b]	0.0	0.0	0.0	0.0	0.0	0.0	-0.1	0.1

[a] see Table 5.5 for definitions. [b] proportion of dam capacity full at start

The strong seasonality of flow and inter-annual variation is shown in Figure 8.4a. In a normal year (e.g. Figure 8.4b time steps 1-60), the upper catchments B74, B75 and B89 peak early, with the gauges below dams (B57 and B51) peaking later in the season as the dams fill. In drought years (e.g. Figure 8.4b time steps 60-130), a bimodal discharge regime can be observed as the dams are emptied to satisfy demand. The alluvial aquifers are filled up by the first flows, and only start to fall towards the end of each season (Figure 8.5a), whereas Insiza Dam falls throughout the year and is only fully replenished during years with higher flows (Figure 8.5a).

It can be seen that demand outstrips supply in the Insiza Catchment, with some shortages recorded in most years (Figure 8.5b). This shortage is most common for demand nodes downstream of Insiza Dam. Such a result is to be expected, given the priority given to water supply for the City of Bulawayo, which draws from Insiza Dam – and the occasional conflict between water users in the lower Insiza (Mutezo, 2008).

8.4.2.   Scenario modelling

The results are summarised in Table 8.9 and the full details are presented in Supplementary Material Table 8.S1.

Two scenarios of expected change were simulated: In the climate change scenario (1), the reduction in rainfall of 10% resulted in an over-proportional reduction of discharge (output from HBVx) of 24% or 6.8 mm a^{-1}. The impact of this within the catchment was split roughly evenly between a 3.14 mm a^{-1} decline in productive use of water (22% fall compared to baseline) and a 3.22 mm a^{-1} decline in outflow from the catchment (32% fall compared to baseline). This scenario only assessed the effect of climate change on rainfall and rainfall-interception-evaporation-runoff partitioning. It does not, for example, show the effect of rainfall reduction on land cover (e.g. Hiernaux et al., 2009), and the concomitant effect on discharge.

Within the limitations discussed above (8.3.6), the model shows that land use change due to rural population growth (3) results in a slight increase in discharge. Inflows later in the season mean that storage in the dams fall less quickly, spills from the dam during the wet season are less and the increased inflow (+0.66 mm a^{-1}) is entirely converted to productive use (also +0.66 mm a^{-1}), decreasing the frequency of shortages slightly.

Three scenarios of possible changes in rainfed farming were simulated: In scenario 4A, the change from maize to small grains, had minimal impact on discharge and catchment outflow. Mupangwa et al. (2012) showed average yields of 2.0 t ha^{-1} in sorghum compared to 1.3 t ha^{-1} in maize under rainfed conditions at Matopos, west of the study area. At these yields, over 12,000 additional tonnes of grain per year could be expected under sorghum production. Given the very low maize yields in Insiza District (0.13 to 0.18 t ha-1, FAO, 2012), and low household cereal production in Matabeleland South (41 kg maize and 13 kg small grains, ZIMVAC, 2012) this would be a substantial food security boost – with minimal impact on water resources.

In scenario 5 the conversion of a quarter of maize fields into fodder production showed a minor drop in discharge (-0.18 mm a^{-1}) resulting in a similar drop in outflow from the catchment (-0.17 mm a^{-1}). Given the critical importance of livestock feed for supplementary feed during drought years (Fewsnet, 2012) and for raising the sale value of livestock (Homann and van Rooyen, 2008), the minimal hydrological impact is important to note.

In scenario 7, the effect of conservation agriculture was simulated, showing a 3% decrease in discharge, leading to a 0.61 mm a^{-1} decrease in outflow from the catchment. Yield gains of 50% to 200% have been suggested by Twomlow et al. (2008), which would result in a gain of 1,700 to 9,500 t a^{-1} (compared to post-harvest survey baseline: FAO, 2012) for the Insiza Catchment. Thus, while conservation agriculture has been widely shown to increase rainfed productivity (Mupangwa et al., 2008) and household income (Mazvimavi and Twomlow, 2009), this model suggests its downstream impact is may be significant.

Conjunctive use of the large reservoirs offers major efficiency improvements to the catchment. In scenario 2, an increase of 0.43 mm a^{-1} in productive use of water (irrigation and mining) is achieved through a 0.21 mm a^{-1} reduction in evaporation losses (9% reduction) and only a 0.15 mm a^{-1} decline in outflow from the catchment (as well as a change in water stored).

Three scenarios on irrigation were simulated: In scenario 6, the additional productive water (8 % increase over the baseline productive use or 2.2% of inflows) is achieved partly through reduced evaporation (0.27 mm a^{-1} or 11% of baseline evaporation losses), but also at the expense of significant decline (0.70 mm a^{-1} or 7 %) in outflow from the catchment. Evaporation losses are also reduced by the abstractions from the aquifers lowering the water table below evaporation extinction depth some of the time. Maintaining frequent recharge of the alluvial aquifer system is also important to avoid water quality problems during dry periods, where abstraction may lead to intrusion of more saline waters (Moyce et al., 2010). The irrigated area could be expanded further (see Supplementary Material Figure 8.S2), but substantial shortages are then experienced.

Scenario 9A considers the conversion of 2,760 ha of irrigated land to drip, which results in savings in water use: a drop of 4% in productive water use, leading to a substantial decrease in the frequency of shortage to water users (20 %) and an increase in outflow from the catchment (0.46 mm a^{-1}). Thus, drip irrigation not only represents a more efficient (Moyo et al., 2006) and economic (Woltering et al., 2011) system at field scale, but can substantially reduce downstream impacts of irrigation at basin scale.

In scenario 4B, the change of crop from irrigated maize to biofuels results in a small increase in productive water use which leads to a 2% decrease in outflows from the catchment. Combining this scenario with changes in rainfed farming gave similar results. Although the land allocation aspects of such change are controversial (Scoones et al., 2012), it is clear that the hydrological impacts at this scale are not substantial.

The simulation of six new small dams (scenario 8A) allows for increased productive use of water (5% above baseline or 0.67 mm a^{-1}) and decrease in shortage to water users (13% lower than in baseline). Operating the dams in conjunctive use (as simulated in scenario 2) means that evaporation losses are reduced (-0.14 mm a^{-1}) but there is still impact on outflow from the basin (-0.45 mm a^{-1}).

Two multiple use system (MUS) scenarios were simulated: In scenario 8B, each existing dam without significant water use and each simulated new dam from scenario 8A, is allocated a small irrigation scheme, the increase in productive water use is 1.06 mm a^{-1} at the expense of 0.52 mm a^{-1} drop in outflow, with evaporation losses dropping by 0.46 mm a^{-1} compared to the baseline. This is a substantial efficiency gain. When the irrigation schemes use drip irrigation (scenario 9B), the efficiency savings are substantial: consumption of water in productive use is only 2% higher than the baseline, compared to 8% higher in conventional irrigation, and the decrease in outflow from the catchment is less than half (0.20 mm a^{-1} vs. 0.52 mm a^{-1}).

8.5. Conclusions and Recommendation

8.5.1. Methodological considerations: HOWSIT model and data-poor catchments

HOWSIT performed adequately on validation. However, the changes simulated in scenario modelling for productive use of water and for outflow from the catchment are below the 8% sensitivity threshold established for a 10% change in some of the soil parameters (section 3.1) – with the exception of the climate change scenario. Simulated changes in evaporative losses were above the threshold for half of the scenarios. Whilst more complex, GIS-based models can give impressive results in data-rich catchments (e.g. Van Ty *et al.*, 2011), in the data-poor Mzingwane Catchment, this type of coarse scenario modelling can be useful for planning purposes, given also the early stage of river basin planning of the Mzingwane Catchment Council (Sakuhuni et al., 2012). The model thus has the potential for use in similar data-poor catchments. Improvements could include the incorporation of a sediment module to WAFLEX, to allow for modeling of erosion and siltation.

8.5.2. Climate change and the Insiza Catchment

The more than proportional response of discharge to a decline in rainfall is well illustrated by the model, as are the substantial water shortages to be anticipated from climate change. The changes in crop type (4A) and irrigation method (9A) discussed above can mitigate the impact on continued agricultural production under rainfed and irrigated management respectively, as can conservation agriculture (S7).

8.5.3. Food production gains and downstream implications

Potentially the largest livelihood gain is fodder cropping for cattle. The 6,601 ha of fodder crops (scenario 5) could produce 12,271 t a^{-1} (Li et al., 2007: 1.927 t ha^{-1}). This is sufficient for the feeding of 16,118 beasts per year (Siemens et al., 1999: 6.6 kg d^{-1}). Even at the Christmastime low price of $ 275 per beast (Fewsnet, 2012), one beast is sufficient to secure the annual cereal needs of a household (764 kg household^{-1} a^{-1}: ZIMVAC, 2012; $0.39 kg^{-1} maize: Fewsnet 2012). Fodder cropping could therefore benefit at least 16,000 households in the Insiza Catchment, without a significant effect on the catchment water balance.

Another substantial livelihood gain would be the switch from maize to small grains. The estimated additional yield of 12,000 t a^{-1} (scenario 4A: yield of 2.0 t ha^{-1} in sorghum compared to 1.3 t ha^{-1} in maize, gives yield increase of 0.7 t ha^{-1} , applied across two-thirds of the cultivated land in the catchment) is sufficient to meet the cereal needs of 16,653 households. This would result in a small decrease in catchment outflow.

The potential for conservation agriculture to upgrade smallholder rainfed farming has been demonstrated by others (e.g. Mupangwa et al., 2007; 2008; Kasam et al., 2009) although with necessary caution on where it may be adopted successfully (e.g. Giller et al., 2009; Ncube et al., 2009). The yield gain of 1,700 to 9,500 (scenario 7) is sufficient for the annual cereal needs of 2,300 to 12,845 households. However, given the low maize yields in Matabeleland South (41 kg a^{-1}: ZIMVAC, 2012), even the highest gains in conservation agriculture still leave average household cereal production at a small fraction of annual cereal needs. Furthermore, the downstream impact is relatively high, with a simulated decrease in outflow of 6%.

Increases in irrigation are possible, with 308 ha from smallholder irrigation from alluvial aquifers (scenario 6) and 229 ha from schemes that could be set up at existing and new reservoirs (scenario 9A), a total increase of 19% on the existing 2,760 ha in the catchment. However, although the benefits of irrigation to individual households are substantial, the number of households is small: 1,078 households could be supplied with irrigation water from alluvial aquifers and 802 households with irrigation water from multiple use of existingand new reservoirs (0.29 ha household^{-1} a^{-1}: Maisiri, et al., 2005). These development scenarios have a downstream impact: using the alluvial aquifer for smallholder irrigation could decrease outflow by 7%. Construction of new reservoirs would decrease outflow by 4.5 % (up to 5.2% with use of water from these reservoirs for irrigation). This is a consideration for large downstream users such as the Zhovhe citrus farms and Beitbridge (see chapter 7).

Thus it can be seen that the largest livelihood gains could potentially be achieved through changes in crops grown in rainfed farming, rather than capital-intensive investments in irrigation infrastructure. However, this may be in part due to the fact that the Insiza Catchment is a headwater catchment with heavily-committed water resources, and greater livelihood gains from irrigation could be expected in lowland areas (e.g. Love et al., 2010c).

8.5.4. *Efficiency gains*

The model has illustrated the important efficiency gains that can be made from conjunctive use of reservoirs (where upstream dams release water when the storage in the dam downstream   falls below a threshold), decreasing both water shortage to users, and also evaporation losses – with dams remaining well below capacity, the smaller surface area decreases evaporation. This is possible even though the upstream storage is distributed – not one single upstream reservoir. This is essential for fairer access across a catchment, although there would be substantial administrative and institutional challenges (Van der Zaag and Gupta, 2008).

The potential for drip irrigation to change the character of water use in the catchment is substantial, with a third of the water shortage eliminated and a 4.6% increased outflow from the catchment.

8.6. Supplementary Material

8.6.1. HOWSIT User Interface

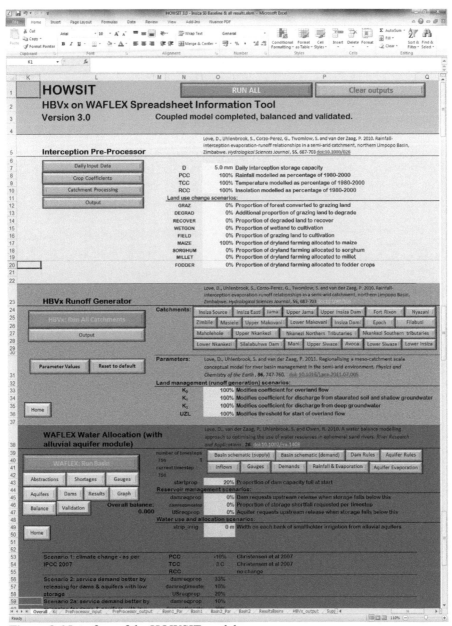

Figure 8.6. Interface of the HOWSIT model.

9. Conclusions and Recommendations

9.1. Conclusions

This chapter synthesises the main findings of previous chapters. Reference is also made to additional publications arising from this study – the abstracts of these publications are presented in Annex 1.

9.1.1. Food security and water availability in semi-arid Zimbabwe

The Millennium Development Goals' target to halve the proportion of people who suffer from hunger is extremely important in southern Africa, where food security has become increasingly problematic over the last 20 years. In Zimbabwe in 2012, 19 % of households are considered food insecure, rising to an average of 30 % in the semi-arid Mzingwane Catchment. The two most insecure districts in the country, Gwanda with 57 % of households insecure and Mangwe with 53 %, are in the Mzingwane Catchment (ZIMVAC, 2012). In this study, through evaluation of scenarios (**Chapters 7 and 8**), we have seen how in water-scarce basins like the Limpopo, while water resource availability imposes constraints on food production, opportunities exist to use water resources more effectively. Many of the challenges to crop and livestock productivity have their origins in water scarcity or water management (see Figure 9.1). Access to water is the community challenge ranked highest in household surveys (ZIMVAC, 2012). Access is limited by actual scarcity, availability and affordability of water storage and appropriate abstraction technology and water allocation practices. Increased knowledge, improved technology transfer and better allocation models and practices can lead to more equitable and productive use of water in the semi-arid lands.

It is clear that many of these strategies to address hunger in southern Africa are water intensive (downstream impact - **Chapters 7 and 8**), and some have been shown to result in water use conflicts (**Chapter 2, Annex 1.4, Annex 1.8**). Such initiatives should be complemented by interventions that maximise the use of the existing scarce water resources, whether in rainfall, or surface storage, or groundwater.

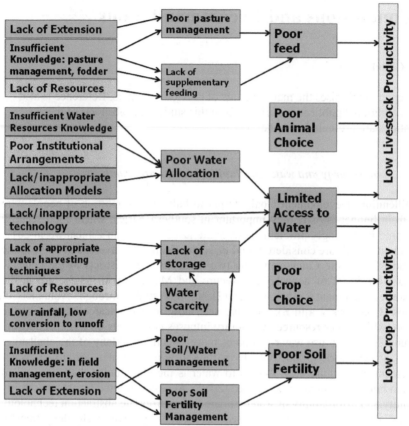

Figure 9.1. Flow chart of major challenges to food production in the semi-arid Limpopo Basin. Adapted from Nyabeze and Love (2007).

9.1.2. *Hydrological characteristics of the Mzingwane Catchment*

Characterisation of water availability and scarcity in the Limpopo Basin is a first step towards improving water management and allocation. However, the heterogeneities in space and time of hydrological and climatic parameters in southern Africa are complex. The strongest signals are the effect of ENSO in droughts (such as 1982/83, 1986/87, 1991/92, 1997/98) (**Chapter 3**). Many statistical tests do not show significant change in hydroclimatic trends over the past century (e.g. Mazvimavi, 2010; Mupangwa et al., 2011). However, in this study trend analysis using the Pettitt test along with Cohen's d statistic, Spearman rank test and Mann-Kendall test, were used to evaluate changes that have taken place in rainfall, runoff and temperature within the semi-arid study area over the second half of the twentieth century. These tests showed that since 1980, water resource availability in the northern Limpopo Basin (Mzingwane Catchment) has declined, both in terms of total annual water available for storage (i.e. declines in annual rainfall, annual runoff) and in terms of the frequency of water availability (i.e. declines in number of rainy days, increases in dry spells, increases in days without flow). The risk posed by the changing hydroclimatic patterns observed are an increased risk of rainfed crop failure, due to an increasing probability of dry spells and decreasing number of rainy days, and an increasing risk to water supply from decreasing annual runoff.

In addition to the past trends reported above, water availability may be limited by the effect of global warming. The reduction in rainfall in the Limpopo Basin due to climate change is expected to be between 10% and 20% below the 1900-1970 averages by 2050 (Milly et al., 2008) or by up to 10% below the 1980-1999 averages by 2099 (Christensen et al., 2007) – although some of the models predict lower rates of decline. This will have a more than proportional impact on runoff, with a 10% decline in rainfall causing a simulated 24% decline in discharge in the Insiza Catchment, a 22% decline in productive use of water and a 32% decline in outflow (**Chapter 8**). Recharge to alluvial aquifers will also decline: estimated at 9 % for the Mushawe River in the east of the Mzingwane Catchment (**Chapter 6**). Dam yields for the major reservoirs in the north of the Mzingwane Catchment, which supply Zimbabwe's second-largest city of Bulawayo by inter-basin transfer, have been declining since 1980 and will decline by a further estimated 34-39 % by 2030 (**Annex 1.16**).

The response of meso-catchments in the study area to rainfall is characterised by high levels of interception, slow infiltration and percolation and moderate to fast overland flow. This suggests that rainwater utilisation could be improve through (i) in-field soil water conservation techniques that increase the rate of infiltration and percolation, and (ii) micro-catchment or runoff farming and supplementary irrigation to capture overland flow from areas adjacent to fields. This is particularly important in the degraded catchments that have faster overland flow.

Interception is a critical process in semi-arid meso-catchments and may account for around 30% of rainfall in a normal year but over 50% of rainfall in a drought year, although this should be confirmed by field measurements as the figure could include some part of soil evaporation (**Chapter 4**). Runoff occurs in discrete events, except for meso-catchments in the moister north-west. Baseflow is higher in granitic catchments than in those with other bedrock lithologies. Land degradation, which is increasing in the study area (Kileshye-Onema and van Rooyen, 2007), has a strong influence on flow processes, increasing rapid, overland flow (**Chapter 5**).

Some significant water quality problems have been identified in the study area, including salinity of some alluvial aquifers and pollution associated with mining (**Annexes 1.15, 1.16**).

9.1.3. *Water and land strategies to increase food production*

Three groups of strategies have been considered in this study:
(i) better land and soil/water management,
(ii) better surface water management, and
(iii) exploitation of alluvial aquifers (conjunctive surface water – groundwater use).

Figure 9.2 presents the major benefit of each strategy, in terms of number of households whose livelihoods could be sustained by the strategy, compared against the impact in terms of reduced downstream flows.

Some interventions target the household scale: For the land-poor, the sick, and often for women, the homestead is often the only site where they can use water productively and in a resilient way (**Annex 1.20**). Household drip kits are an important example of major potential improvements in productive use of water (**Annex 1.4**). However, due to the variety and distributed nature of water sources for household use, the downstream impact of such interventions is difficult to evaluate.

Land and soil/water management

It can be seen that the greatest benefit for the least impact comes from the strategies providing for better land and soil/water management. Changing crop choice and fodder production show clearly excellent results, with conservation agriculture having potentially a similar impact. Switching dryland cropping from maize to small grains has long been an objective of agricultural policy and extension in Zimbabwe (e.g. Mupangwa *et al.*, 2011) and the impact in terms of increased cereal production should be impressive. Replacing maize with fodder crops also allows for a substantial livelihood gain, as sales of meat can fund the purchase of far more cereal than could be produced from the same land (**Chapter 8**). Conservation agriculture has potentially a high impact in this basin (e.g. Mupangwa *et al.*, 2007, 2008) although the yield gains are variable and careful assessment is needed prior to successful adoption (e.g. Giller *et al.*, 2009). Rainwater harvesting is a key technology in this context (**Annex 1.9**), although in this study it was only possible to investigate the downstream impact of in-field rainwater harvesting techniques; ex-field rainwater harvesting was not modelled.

The high benefit/impact ratio in the land and soil/water management is partly due to the fact that these are rainfed farming strategies, and rainfed farmland occupies a very large area of land compared to irrigation – and obviously uses less water. It should also be considered that whilst these strategies do not require much investment in agricultural or water supply infrastructure, considerable investment is needed in market linkages, extension and political will.

Surface water management

One of the most common interventions by government and NGOs is construction of small dams. Small dams are a key resource to rural communities, and can be readily managed at village level (**Annex 1.18**). It has been shown in this study and elsewhere (e.g. Volta Basin: Andreini *et al.*, 2009) that evaporation losses from such dams are relatively small: 6 % of total storage, although considerably higher than amounts consumed in livestock watering, less than 1 % (**Chapter 7**). The relatively small amounts of water used in dams built for livestock allows for a conversion to Multiple Use Systems through abstraction of water for irrigation. In the Insiza Catchment this could provide 229 ha of irrigation for 802 households, although the downstream impact is relatively large for this benefit, unless drip irrigation is practiced (**Chapter 8**). The reduced downstream impact from drip irrigation as compared to conventional irrigation can be seen in Figure 9.2.

Some of the large dams in the Mzingwane Catchment are under-utilised. Better management of these dams, together with conjunctive reservoir use (where water is released from an upstream reservoir, when the water level in a downstream reservoir falls below a critical level) could allow for 1,250 ha irrigation for 4,375 households (**Chapter 7**). This is a significant livelihood gain without the need for construction of a new reservoir.

In contrast, construction of new large reservoirs for irrigation has considerable downstream impact, in addition to the costs of the infrastructure. The changing downstream discharge pattern, especially the loss of low flows, has been demonstrated in the Insiza Catchment (**Annex 1.2**). Low flows disappear completely downstream of Zhovhe Dam in the Mzingwane Catchment, and there is also loss of alluvial material and recolonisation of the riverbed by vegetation (**Chapter 7: Section 7.6**). Declining water resources availability and the impact of climate change also suggest that some of these planned developments may not be as sustainable or have as high a yield as originally thought. Examples (see Figure 1.1 for locations) include Glassblock Dam on the Mzingwane River (**Chapter 3**) and the Elliot and Moswa Dams on the Thuli River (**Annex 1.1**).

Conjunctive surface water – groundwater use

There is great potential for the exploitation of alluvial aquifers for irrigation water supply. A total of 6,738 ha of land can be irrigated by exploitation of alluvial aquifers, mainly in the lowland rivers and tributaries. This is discussed in more detail in Section 9.1.4 below.

9.1.4. The potential of alluvial groundwater

Storage and flow in small alluvial aquifers – the sand beds of ephemeral rivers in meso-catchments – are dominated by recharge from the river on an event basis, and also by seepage and evaporation, where these processes occur (**Chapter 6**). However, when the water table falls below evaporation extinction depth (about 0.9 m; Nord, 1985) and where seepage is minimised by an impermeable basement, water levels can be nearly static. Thus the primary limitations to alluvial aquifer viability are (i) scale, as smaller catchments have smaller river beds thus less storage below evaporation extinction depth; and (ii) geology, which controls basement permeability, as well as the composition of the alluvial sediment (**Annexes 1.6, 1.12**).

In this thesis I have distinguished small (chapter 6) and large (chapter 7) alluvial aquifers. Small alluvial aquifers of this scale are likely to be suitable for use at the scale of domestic and livestock water supply and the irrigation of small gardens (sub-hectare to several hectares per km of river reach). Such gardens would be best suited for horticulture, to maximise the value obtained from the irrigation water, and are replicable on many river reaches. The small alluvial aquifers offer the potential for distributed, localised water storage, readily accessible to a large number of communities with limited financial resources. This small-scale technology is both appropriate for alluvial aquifers in small sand rivers and also more likely to be cost-effective and within the reach of smallholder farming communities. In the Mzingwane Catchment, upto 2,800 ha of such aquifers could be exploited to provide irrigation for some 9,800 households (**Chapter 6**).

Storage and flow in large alluvial aquifers – the sand beds of major lowland rivers – are dominated mainly by recharge from the river, whether natural or from managed releases from reservoirs (**Chapter 7**). The great depth of the sand beds minimises the effect of evaporation on the water balance and groundwater flow downstream in the aquifer generally exceeds evaporation by a factor of at least 10, recharging the downstream aquifer blocks and allowing for abstractions from these blocks during the dry season. Alluvium in the lower Mzingwane, Thuli and Shashe rivers forms ribbon shaped aquifers extending along the channel and reaching over 20 km in length in some localities. These alluvial aquifers extend laterally outside the active channel (**Annex 1.1**).

On these large rivers, alluvial aquifers could be exploited to irrigate strips of land along each bank of the river, provided suitable soils are available (**Chapter 7**). Where conditions are favourable, such as at geological boundaries which hold back the groundwater, the area irrigable would be larger (**Annex 1.13**). This irrigation would be decentralised, owned and operated at household level and the benefits would have the potential to reach a much larger proportion of the population than is currently served. Demand is most easily met through the use of upstream dams to resupply the alluvial aquifers, whilst still supplying their existing users. This can maximise productive use of water compared to evaporation losses from dam and aquifer storage. Such conjunctive use of surface water and groundwater is more efficient than the traditional approach where alluvial aquifer recharge is considered as "transmission losses". Modelling in this study suggests the potential to irrigate 3,630 ha on the lower Mzingwane River (**Chapter 7**) and 308 ha on the lower Insiza River (**Chapter 8**). The frequent releases from upstream dams would minimise the intrusion into the alluvial aquifer in the lower Mzingwane of slightly saline water from the flood plain aquifers, which is a problem for irrigation water supply especially during dry years (**Annex 1.1**).

9.1.5. *Innovations in water resources modelling and their application*

HBVx Rainfall-runoff-interception-evaporation model

The HBVx model, adapted from HBV Light (Seibert, 2002) with the incorporation of an interception pre-processor, satisfactorily models the ephemeral surface flow and the minimal baseflow from deep groundwater in semi-arid meso-catchments (**Chapter 4**). Regionalisation of the model showed inadequacies in performance at the daily time step, likely due to the very high spatial variability of rainfall at this time step. The best-performing parameter sets produced mainly negative volume errors, which are conservative for water resource modelling and water allocation but problematic for flood prediction. Given these two factors, the HBVx model is probably more useful in the study area for providing input to water resources modelling and allocation planning than for detailed process hydrology (**Chapter 5**).

WAFLEX water resources model

The spreadsheet-based model WAFLEX (Savenije, 1995) is a useful tool for assessing existing upstream-downstream interactions and future development and allocation options (**Annex 1.8**). In this study, WAFLEX was adapted by the incorporation of an alluvial groundwater module, simulating groundwater storage, groundwater evaporation, abstraction, recharge and Darcian flow in alluvial groundwater blocks. The model performed adequately and produced a comparable trend in findings on alluvial aquifer behaviour to published field studies. It thus provides a flexible tool for the evaluation of alluvial aquifers on large, lowland rivers and can provide useful information for planning purposes from limited data (**Chapter 7**).

HOWSIT coupled model

Spreadsheet based models like HBVx and WAFLEX have the ability to provide useful planning information from limited data. Coupling the two models in HOWSIT provides a comprehensive and integrated water resources model for semi-arid catchments. The coupled HOWSIT model performs moderately well in this context, but shows significant sensitivity to soil parameters, higher than the change simulated in many development scenarios. Model performance was constrained by the high processes variability of the semi-arid environment, especially soil processes. It is thus likely that the model could perform better in catchments with greater availability of soil data (**Chapter 8**).

Risk in water resources and allocation planning can be decreased through the use of conservative assumptions, with which HOWSIT can give robust minimum values for water allocation purposes. The coupled model could thus assist in drought risk management (**Chapter 8**).

Classification of alluvial aquifers by remote sensing

On the false colour composite band 3, band 4 and band 5, alluvial deposits stand out as white, and dense actively growing vegetation stands out as green making it possible to mark out the lateral extent of the saturated alluvial plain deposits using the riverine fringe vegetation signature (**Annex 1.1**). This allows for rapid identification of alluvial aquifers. A classification using existing regional information on the basement geology and erosion surface can assist in targeting favourable reaches of these sand rivers for further evaluation and possible water supply development (**Chapter 6**).

9.2.   Recommendations

9.2.1.   *Water Resources Development*

Future large scale water resources development should consider the changing water resources availability in the Limpopo Basin and the impact of climate change. The likely yield of proposed reservoirs such as Glassblock, Oakley Block, Moswa and Elliot should be re-assessed prior to investment, as well as the downstream impact of abstractions and evaporation losses from these reservoirs. A thorough investigation of possible water supply from alluvial aquifers should be part of the scoping exercise when considering the possible construction of these or other reservoirs. Further stress to urban water supplies in the study area, notably those of Zimbabwe's second-largest city of Bulawayo, which already experiences chronic water shortages

(Gumbo, 2004), should be expected. Water resource development should consider alternatives to the Mzingwane Catchment and the Limpopo Basin, such as the long-awaited Matabeleland-Zambezi water carrier.

Despite the challenges, there remains scope for expansion of irrigation in the study area, which is the livelihood intervention given highest priority by government. The total additional irrigation potential suggested by this study is 8,200 ha. This could provide smallholder irrigation for close to 29,000 households.

Most of the additional irrigation scope (82%) comes from potential exploitation of alluvial groundwater. Functional and operational decentralisation of smallholder irrigation, using strips of land adjacent to rivers rather than concentrating only on large schemes, should allow for low cost abstraction systems and household scale control of water use, leading to better household investment decisions and improved access to water for women. However, it should be noted that this type of development has significant downstream impact, with use of the alluvial aquifer for smallholder irrigation decreasing simulated outflow by 7% in the Insiza Catchment and 20% in the Lower Mzingwane.

9.2.2. Land and Agricultural Water Management

Priority should be given in extension and outreach to changes in rainfed cropping which can benefit a greater proportion of the population without significant downstream impact on water resources. This should include changing crop from maize to small grains and the production of cattle fodder.

At the field scale, the dominance of slow infiltration and percolation and moderate to fast overland flow suggests that rainfed agricultural water management would benefit from techniques to increase infiltration and percolation, such as mulching, terracing and deep-winter ploughing, and rainwater harvesting techniques, both in-field and ex-field, to capture more of the rapid overland flow. Such techniques do have a significant downstream impact, however, with a 6% decline in outflow in the Insiza Catchment simulated.

9.2.3. Water Resources Management

Community or catchment water resource assessments must become an essential precursor to food security interventions, due to the convergence of water scarcity and food scarcity, and the constraints that water resource availability impose on development initiatives in basins such as the Limpopo, in order to ascertain that the water supply assumed for a project is sufficient, sustainable and acceptable to the community, and to decrease the risk of conflicts between water users.

It is particularly important that water-saving and efficiency gain techniques be used to maximise the use of the existing scarce water resources, whether rainfall, surface water or groundwater. The greatest water efficiency gains could be achieved through:

    (i)   Soil water conservation (**Annex 1.3**)

    (ii)   Conjunctive use of surface water and groundwater, and conjunctive use of downstream and upstream dams (where upstream dams release water when the storage in the dam downstream falls below a threshold). This decreases both water shortage and evaporation losses.

(iii)   Change of irrigation method to drip, with catchment water shortage reduced by one third in simulations of the Insiza Catchment.

Investment in the first two approaches requires limited additional infrastructure, and thus is likely substantially cheaper than the construction of new reservoirs - and yet at the same time leads to greater increases in water availability for productive use. However, it should be noted that such increases in water availability would likely be smaller in basins which are closing, or over-exploited.

These approaches need to be built into the extension curriculum and the training of catchment councillors. Increased knowledge, improved technology transfer and better allocation models and practices can lead to more equitable and productive use of water in the semi-arid lands.

9.2.4.   *Future Research Directions*

Data availability and density remains a huge challenge in the study area. Without more reliable rainfall and runoff data with longer time series, modelling hydrological processes in semi-arid ephemeral catchments will remain difficult. For rainfall and evaporation data, combining remotely-sensed data with ground stations may be the way forward. Soil data is particularly important for process modelling – and particularly scarce. Remote sensing using thermal imaging for soil water content (e.g. Ahmad *et al.*, 2010; Mulder *et al.*, 2011), and ground-calibrated NDVI for other characteristics (e.g. Dang *et al.*, 2011) are good possibilities. This would allow for more detailed studies of the tributary meso-catchments in the Mzingwane Catchment than was possible in this study (**Chapters 4, 5**).

The declining trends of water availability in the Mzingwane Catchment identified in this study (**Chapter 3**) should be compared with similar analyses that could be done of data from other riparian states of the Limpopo Basin, to provide information for planning at basin scale.

Erosion and siltation are a major problem due to land degradation (**Chapter 5, Section 8.3.6**) and activities such as gold panning (**Annex 1.14**). The incorporation of a sediment module to WAFLEX could allow for modeling of erosion and siltation and its downstream effects.

The modified WAFLEX approach to modelling lowland alluvial aquifers in ephemeral rivers (**Chapter 7**) can be applied to evaluate the potential of alluvial aquifers elsewhere in semi-arid Africa, especially in water-scarce countries such as Botswana and Namibia.

Detailed studies of specific alluvial groundwater sites of high potential need to be done in order to facilitate water resources development. This has already been done at Malala in the Lower Mzingwane (**Annex 1.13**) but should be done at other sites identified in this study (**Chapters 6, 7**). The remote sensing approach to identification and targeting of alluvial aquifers (**Chapter 6**) should be tested and ground-truthed in other basins.

Samenvatting

WATERBEHEER STRATEGIEËN DIE DE VOEDSELPRODUCTIE IN DE SEMI ARIDE TROPEN VERHOGEN
MET SPECIALE NADRUK OP DE MOGELIJKHEDEN VAN ALLUVIAAL GRONDWATER

Een aantal hydro-klimatologische en institutionele factoren onderstrepen het belang van het investeren in waterbeheer en het modelleren van waterbeheer in zuidelijk Afrika. De vraag naar water stijgt voortdurend vanwege uitbreidende stedelijke gebieden en grotere behoefte voor water voor de landbouw vanwege het bereiken van de voedselzekerheid millenniumdoelstelling. In doorsnee jaren is de waterbehoefte (voornamelijk van de landbouw en stedelijke gebieden) in precair evenwicht met het beschikbare water, terwijl er bij droogte grote tekorten ontstaan met voedselonzekerheid als gevolg. Toegang tot water wordt beperkt door absolute schaarste, door beschikbaarheid en betaalbaarheid van wateropslag, door passende onttrekkingtechnologieën en door waterverdelingpraktijken. Deze studie toont aan dat de beschikbaarheid van waterbronnen in de noordelijke Limpopo rivierbekken (d.w.z. het deel gelegen in Zimbabwe, ook wel bekend als het Mzingwane Stroomgebied) zijn afgenomen in de laatste 30 jaar, zowel in termen van de totale jaarlijkse waterbeschikbaarheid voor opslag (d.w.z. afname in jaarlijkse neerslag en jaarlijkse afvoer) en in termen van de frequentie van waterbeschikbaarheid (d.w.z. afname van het aantal regendagen, toename van droogteperiodes en van dagen zonder rivierafvoer). Bovendien voorspellen verscheidene klimaatmodellen dat zuidelijk Afrika in de komende vijftig jaar aanzienlijk minder regenval en oppervlakkige afstroming zal ondervinden. Simulatiemodellen suggereren een meer dan evenredige afname in regenval en beschikbaar water voor productief gebruik.

Veranderingen in strategieën van land- en watergebruik kunnen significante effecten hebben op de waterhuishouding. Uitbreiding van irrigatie en de constructie van reservoirs hebben vanzelfsprekende effecten op riviersystemen, maar regenafhankelijke landbouw en veranderingen van landgebruik, zelfs als deze geen aanspraak maken op grond- of oppervlaktewater, kunnen van sterke invloed zijn op het afstromingproces. In deze context is er een duidelijke behoefte aan het modelleren van waterstromen om integraalwaterbeheer en planning te kunnen ondersteunen, zodat er een balans gevonden kan worden in het verdelen van water voor voedselzekerheid, andere economische behoeften en de behoeften van het milieu.

Deze studie wil watervoorraden modelleren op schaal van een stroomgebied zodat het effect van verschillende land- en watergebruikstrategieën bij verschillende hydrologische en klimatologische omstandigheden kan worden gekwantificeerd.

Een uitbreiding van het 'HBV light' regen-afvoer model werd ontwikkeld (aangeduid als HBVx), waarbij een interceptie opslag werd geïntroduceerd en waarbij alle berekeningsroutines worden uitgevoerd in semi-gedistribueerde modus via Visual Basic macros in een spreadsheet. Dit werd gebruikt om de respons van regenval in meso-stroomgebieden in het studiegebied te karakteriseren, in termen van de productie van afvoer in relatie tot interceptie, transpiratie en verdamping van water. Dit is belangrijk in kleine semi-aride stroomgebieden, waar sporadische, heftige regenbuien vaak het merendeel van de seizoensafvoer uitmaken. HBVx is geregionaliseerd over 19 meso-stroomgebieden en kan efemere oppervlakteafvoer en minimale basisafvoer afkomstig van diep grondwater in semi-aride meso-stroomgebieden modelleren. Meso-stroomgebieden in het studiegebied worden gekarakteriseerd door hoge niveaus van interceptie, langzame infiltratie en percolatie en matig tot snelle oppervlakkige afstroming.

In efemere rivieren in semi-aride gebieden drogen de alluviale watervoerende lagen die het bed van de zandrivieren vormen niet op, het water verdampt er slechts gedeeltelijk en is vaak van goede kwaliteit. Het noordelijke Limpopo stroomgebied heeft onregelmatige en onbetrouwbare regenval en een zeer lage gemiddelde jaarlijkse afvoer.

Alluviale watervoerende lagen vertegenwoordigen zo een aantrekkelijke optie voor waterbeheer: ten eerste gecombineerd met oppervlaktewater voor de opslag van water in efemere zandrivieren en ten tweede als duurzaam alternatief voor het gebruik van oppervlaktewater. Het water potentieel van een bestudeerde alluviale watervoerende laag werd geëvalueerd met behulp van veldwaarnemingen en het eindige-differentie grondwaterstromingsmodel MODFLOW. Het gedrag van de watervoerende laag bij hogere percolatie, klimaatverandering en ontwikkelingsscenarios werden ook gemodelleerd. Hieruit bleek dat alluviale watervoerende lagen op deze schaal geschikt zijn als watervoorziening voor huishoudelijk gebruik, irrigatie van kleine tuinen en als watervoorziening voor vee.

Een remote-sensing benadering werd gebruikt om 1.835 km alluviale watervoerende lagen te identificeren en in kaart te brengen in het noordelijke Limpopo stroomgebied.

Door waarnemingen in het veld, in het laboratorium te combineren met remote sensing en andere data kon het benedenstroomse deel van de Mzingwane vallei met behulp van het WAFLEX spreadsheet model en een nieuw ingebouwde module, die de water balans uitrekent van in blokken gegroepeerde alluviale watervoerende lagen, succesvol gemodelleerd worden. Het model presteerde afdoende, en genereerde trends in het gedrag van alluviale watervoerende lagen die vergelijkbaar zijn met gepubliceerde veldstudies. Het blijkt dus een flexibel instrument om alluviale watervoerende lagen op grote rivieren te evalueren, en kan zelfs met beperkte data waardevolle informatie genereren voor planningdoeleinden.

HOWSIT is een gekoppeld, spreadsheet model dat als evaluatie-instrument gebruikt wordt voor land en water scenarios en interventiestrategieën gericht op het verbeteren van het levensonderhoud van huishoudens in het Insiza stroomgebied in het Limpopo bekken. Het model integreert HBVx en WAFLEX in een spreadsheet en evalueert klimaatverandering en de respons van interventies zoals het veranderen van gewassoort, irrigatiemethode, duurzame regenafhankelijke landbouw, nieuwe geïrrigeerde stukken land, gewas-vee integratie, op de bodem- en waterdynamica, afvoer, aanvullen van de watervoerende laag en de waterverdeling. Het gekoppelde HOWSIT model presteert in deze context redelijk goed, maar is zeer gevoelig voor bodemparameters; hoger dan de verandering gesimuleerd in veel ontwikkelingsscenarios. Onzekerheden in de planning van waterbeheer water verdeling kunnen worden verminderd door het gebruik van voorzichtige aannames, en zo kan HOWSIT degelijke minimumwaardes produceren voor waterverdelingsvraagstukken.

Strategieën die gericht zijn op het verbeteren van het land en waterbeheer hebben de grootste voordelen bij de minste impact. Het overstappen van maïs naar kleine granen en de productie van veevoer zijn maatregelen die uitstekende resultaten geven. Duurzame regenafhankelijke landbouw heeft mogelijk een vergelijkbaar effect. Kleine dammen zijn een belangrijke waterbron voor plattelandsgemeenschappen, en een overstap naar Meervoudige Gebruik Systemen door het gebruik voor irrigatiewater toe te voegen heeft duidelijke voordelen op lokaal niveau, maar heeft stroomafwaarts een grote impact tenzij druppel irrigatie wordt gebruikt. Verbeterd beheer van bestaande grote dammen, samen met gecoördineerd reservoir beheer (waarbij bovenstrooms water wordt doorgelaten naar een benedenstrooms reservoir indien waterniveaus onder kritieke waarden dalen) kan het productief watergebruik vergroten en de levensomstandigheden van mensen verbeteren zonder dat het nodig is nieuwe reservoirs te bouwen.

De exploitatie van alluviale watervoerende lagen voor irrigatie heeft een groot potentieel; in totaal kan op deze manier 6.740 hectare land worden geïrrigeerd, vooral in de laaglanden en langs zijrivieren. Deze irrigatie zou gedecentraliseerd zijn, in bezit en beheerd op het huishoudelijk niveau. De voordelen hiervan zouden potentieel een veel groter deel van de populatie kunnen bereiken dan nu het geval is.

Deze studie wordt afgesloten met een reeks aanbevelingen wat betreft het ontsluiten van waterbronnen en het waterbeheer in de landbouw. Cruciaal is dat de uitbreiding van irrigatie in de Mzingwane Stroomgebied zich vooral richt op het gecombineerd gebruik van oppervlakte- en grondwater. Ook moet serieus onderzoek gedaan worden naar de alluviale watervoerende lagen als een mogelijke alternatieve waterbron indien de bouw van nieuwe grote reservoirs wordt overwogen. Functionele en operationele decentralisatie van kleinschalige irrigatie, door bijvoorbeeld stroken land grenzend aan rivieren te gebruiken in plaats van grote centraal beheerde projecten, zou goedkope onttrekkingsystemen en waterbeheer op huishoudelijk niveau mogelijk moeten maken, welke kunnen leiden tot betere investeringsbeslissingen in het huishouden en verbeterde toegang tot water voor vrouwen.

In de landbouwvoorlichting en bewustwordingscampagnes moet de aandacht voor regenafhankelijke gewassen moet prioriteit krijgen omdat daar een groter deel van de samenleving baat bij kan hebben zonder aanmerkelijke benedenstroomse gevolgen in de watervoorziening. Dit betekent onder meer dat men moet overstappen van maïs naar droogte resistente gewassen zoals sorghum and gierst, en de productie van veevoer.

Benaderingen die het voorgestelde gecombineerd gebruik van grond- en oppervlaktewater, samen met veranderingen in regenafhankelijke gewassen toepassen, moeten ingebouwd worden in de curricula van de landbouwvoorlichtingsdienst en in de training van de gekozen vertegenwoordigers van watergebruikers in de waterschappen. Vergrote kennis, verbeterde overdracht van technologie en betere waterverdelingmodellen en praktijken kunnen leiden tot rechtvaardiger en productiever watergebruik in semi-aride gebieden.

References

Abelin, H.; Birgersson, L.; Gidlund, J.; Neretnieks, I. 1991. A large-scale flow and tracer experiment in granite, 1 experimental design and flow distribution. *Water Resources Research*, 27, 3107-3117

Adams, M.; Sibanda, S.; Turner, S. 1999. Land tenure reform and rural livelihoods in southern Africa. *Natural resource perspectives*, **39**, 6-22.

Adekalu, K. O; Okunade, D. A; Osunbitan, J. A. 2006. Compaction and mulching effects on soil loss and runoff from two southwestern Nigeria agricultural soils. *Geoderma*, 137, 226-230.

Aerts, J.; Lasage, R.; Beets, W.; de Moel, H.; Mutiso, G.; de Vries, A. 2007. Robustness of sand storage dams under climate change. *Vadose Zone Journal*, 6, 572-580

Agyare, W.A.; Kyei-Baffour, N.; Ayariga, R.; Gyatsi, K.O.; Barry, B.; Ofori, E. 2009. Irrigation options in the upper east region of Ghana In: *Proceedings of the Workshop on Increasing the Productivity and Sustainability of Rainfed Cropping Systems of Poor, Smallholder Farmers, Tamale, Ghana, 22-25 September 2008*, ed., Humphreys, L., The CGIAR Challenge Program on Water and Food, Colombo. pp259-268.

Ahmad, S.; Kalra, A.; Stephen, H. 2010. Estimating soil moisture using remote sensing data: A machine learning approach. *Advances in Water Resources*, 33, 69-80.

Alam, N.; Olsthoorn, T.N. 2011. Sustainable conjunctive use of surface and ground water: modelling on the basin scale. *Ecopersia* 1, 1-12.

Allan, J.D.; Castillo, M.M. 2007. *Stream Ecology: Structure and Function of Running Waters*. Springer: Berlin, 436p.

Allen, R.G.; Pereira, L.S.; Raes, D.; Smith, M. 1998. Crop evapotranspiration - Guidelines for computing crop water requirements - *FAO Irrigation and Drainage Paper* **56**. Food and Agriculture Organization of the United Nations, Rome.

Alemaw, B.F.; Chaoka, T.R. 2006. The 1950-1998 warm ENSO events and regional implication to river flow variability in Southern Africa. *Water SA*, 32, 459-463.

Anderson, I.P.; Brinn, P.J.; Moyo, M.; Nyamwanza, B. 1993. Physical Resource Inventory of Communal Lands of Zimbabwe. *Natural Resources Institute Bulletin* 60, London.

Anderson M.l.; Kavvas, M.L.; Mierzwa, M.D. 2001. Probabilistic/ensemble forecasting: a case study using hydrologic response distributions associated with El Niño – Southern Oscillation (ENSO). *Journal of Hydrology*, 249, 134-147.

Andersson, L.; Samuelsson, P.; Kjellström, E. 2011. Assessment of climate change impact on water resources in the Pungwe river basin. *Tellus*, 63A, 138-157.

Andreini, M.; Schutze, T.; Senzanje, A.; Rodriguez, L.; Andah, W.; Cecchi, P.; Bolee, E.; van de Giesen, N.; Kemp-Benedikt, E.; Liebe, J. 2009. Small Multi-Purpose Reservoir Ensemble Planning. *Challenge Program on Water and Food Project Report* 46, 55p.

Archer, E.; Mukhala, E.; Walker, S.; Dilley, M.; Masamvu, K. 2007. Sustaining agricultural production and food security in Southern Africa: an improved role for climate prediction? *Climate Change*, 83, 287-300.

Arias-Hidalgo, M.; Bhattacharya, B.; Mynett, A.E., van Griensven, A. 2012. Experiences in using the TRMM data to complement rain gauge data in the

Ecuadorian coastal foothills. *Hydrology and Earth Systems Science Discussions* 9, 12435-12461.

Arnell, N. 2003. Effects of IPCC SRES emissions scenarios on river runoff: a global perspective. *Hydrology and Earth System Sciences*, 7, 619-641.

Ashagrie, A.G.; De Laat, P.J.M.; De Wit, M.J.M.; Tu, M.; Uhlenbrook, S. 2006. Detecting the influence of land use changes on discharges and floods in the Meuse River Basin – the predictive power of a ninety-year rainfall-runoff relation? *Hydrology and Earth System Sciences*, 10, 691-701.

Ashton, P. J.; Love, D.; Mahachi, H.; Dirks, P. 2001. *An Overview of the Impact of Mining and Mineral Processing Operations on Water Resources and Water Quality in the Zambezi, Limpopo and Olifants Catchments in Southern Africa.* CSIR report to the Minerals, Mining and Sustainable Development Project, Southern Africa, 338p. http://pubs.iied.org/pdfs/G00599.pdf

Barker, R.; Molle, F. 2004. Evolution of irrigation in South and Southeast Asia. *Comprehensive Assessment Research Report* 5, International Water Management Institute: Colombo, 55p.

Bangira, C.; Manyevere, A.; 2009. Baseline Report on the Soils of the Limpopo River Basin, a contribution to the Challenge Program on Water and Food Project 17 "Integrated Water Resource Management for Improved Rural Livelihoods: Managing risk, mitigating drought and improving water productivity in the water scarce Limpopo Basin". *WaterNet Working Paper 8*. WaterNet, Harare. http://www.waternetonline.ihe.nl/workingpapers/WP8 Limpopo Soils.pdf

Bari, M.; Smettkem, K.R.J. 2004. Modelling monthly runoff generation processes following land use changes: groundwater-surface runoff interactions. *Hydrology and Earth System Sciences*, 8, 903-922.

Barlow, P.M.; Ahlfeld, D.P.; Dickerman, D.C. 2003. Conjunctive-management models for sustained yield of stream-aquifer systems. *Journal of Water Resources Planning and Management*, 129, 35-48

Basima Busane, Sawunyama, T.;L.; Chinoda, C.; Twikirize, D.; Love, D.; Senzanje, A.; Hoko, Z.; Manzungu, E.; Mangeya, P.; Matura, N.; Mhizha, A.; Sithole, P. 2005. An integrated evaluation of a small reservoir and its contribution to improved rural livelihoods: Sibasa Dam, Limpopo Basin, Zimbabwe. In: *Abstract volume, 6th WaterNet/WARFSA/GWP-SA Symposium, Swaziland, November 2005,* p32. http://www.waternetonline.ihe.nl/challengeprogram/P02%20Sawunyama%20sm all%20dam.pdf **[Annex 1.18 of this thesis]**

Basson, M.S., Rossouw, J.D. 2003. *Limpopo Water Management Area: overview of water resources availability and utilisation.* Report P WMA 04/000/00/0203, Department of Water Affairs and Forestry, Pretoria, South Africa, 55p.

Bear, J. 1972. *Dynamics of Fluids in Porous Media*, Dover Publications, Mineola

Benito, G.; Rohde, R.; Seely, M.; Külls, C.; Dahan, O.; Enzel, Y.; Todd, S.; Botero, B.; Morin, E.; Grodeck, T.; Roberts, C. 2010. Management of alluvial aquifers in two southern African ephemeral rivers: implications for IWRM. *Water Resources Management*, 24, 641-667.

Bennie, A.T.P.; Hensley, M. 2001. Maximizing precipitation utilization in dryland agriculture in South Africa - a review. *Journal of Hydrology*, 241, 124-139.

Bergström, S.; 1992. The HBV model - its structure and applications, SMHI, RH, 4, Norrköping, Sweden.

Beven, K. 2002. Runoff generation in semi-arid areas. In: *Dryland Rivers: Hydrology and Geomorphology of Semi-Arid Channels*, eds., Bull, L.J. and Kirkby, M.J.; Wiley, Chichester, pp57-106.

Blöschl, G.; Sivapalan, M. 1995. Scale issues in hydrological modelling: a review. *Hydrological Processes*, **9**, 251-290.

Bormann, H.; Diekkrüger, B.; 2003. Possibilities and limitations of regional hydrological models applied within an environmental change study in Benin (West Africa). *Physics and Chemistry of the Earth*, 28, 1323-1332.

Bornette, H.; Heiler, G. 1994. Environmental and biological responses of former channels to river incision: A diachronic study on the upper Rhône River. *Regulated Rivers: Research and Management* **9**: 79-92. DOI: 10.1002/rrr.3450090202.

Boroto, R.A. 2001. Limpopo River: steps towards sustainable and integrated water resources management. In: *IAHS Publication 268: Regional Management of Water Resources*, eds., Schumann, A.H.; Acreman, M.C.; Davis, R.; Mariño, M.A.; Rosbjerg, D.; Jun, X., IAHS Press, London, pp33-39.

Boroto, R.A.; Görgens, A.H.M. 2003. Estimating transmission losses along the Limpopo River: an overview of alternative methods. In: *IAHS Publication 278: Hydrology in Mediterranean and Semiarid Regions*, eds., Servat, E.; Najem, W.; Leduc, C.; Shadeel, A., IAHS Press, London, pp138-143.

Brown, C.; King, J. 2003. Environmental flows: concepts and methods. *Water Resources and Environment Technical Note* C1 World Bank, Washington D.C.

Bullock, A. 1992. The role of dambos in determining river flow regimes in Zimbabwe. *Journal of Hydrology* 134, 349-372.

Bullock, A.; McCartney, M.P. 1995. Wetland and river flow interactions in Zimbabwe. In: *L'hydrologie tropicale: géoscience et outil pour le développement (Actes de la conférence de Paris, Mai 1995);* IAHS publication No. 238; eds. Chevallier, P.; Pouyaud, B.; pp305-321.

Burt, R; Wilson, M.A; Kanyanda, C.W; Spurway, J.K.R; Metzler, J.D. 2001. Properties and effects of management on selected granitic soils in Zimbabwe. *Geoderma*, 101, 119-141.

Butterworth, J.A.; MacDonald, D.M.J.; Bromley, J.; Simmonds, L.P.; Lovell, C.J.; Mugabe, F. 1999. Hydrological processes and water resources management in a dryland environment III: groundwater recharge and recession in a shallow weathered aquifer. *Hydrology and Earth Systems Sciences*, 3, 345-352.

Calder, I.R., Hall, R.L., Bastable, H.G., Gunston, H.M., Shela, O., Chirwa, A., Kafundu, R. 1995. The impact of land use change on water resources in sub-Saharan Africa: a modelling study of Lake Malawi. *Journal of Hydrology*, 170, 123-135.

Cane, M. A; Eshel, G; Buckland, R.W. 1994. Forecasting Zimbabwean maize yield using eastern equatorial Pacific seas surface temperatures. *Nature*, 370, 204-205.

Cantón, Y.; Domingo, F.; Solé-Benet, A.; Puigdefábregas, J. 2001. Hydrological and erosion response of a badlands system in semiarid SE Spain. *Journal of Hydrology*, 252, 65-84.

Casenave, A.; Valentin, C. 1992. A runoff capability classification system based on surface features criteria in semi-arid areas of West Africa. *Journal of Hydrology*, 130, 231-249.

Castro, N.M. dos R.; Auzet, A.-V.; Chevallier, P.; Leprun, J.-C. 1999. Land use change effects on runoff and erosion from plot to catchment scale on the basaltic plateau of Southern Brazil. *Hydrological Processes*, 13, 1621-1628.

Cerdan, O.; Le Bissonnais, Y.; Govers, G.; Leconte, V.; van Oost, K.; Couturier, A.; King, C.; Dubreuil, N. 2004. Scale effects on runoff from experimental plots to catchments in agricultural areas in Normandy. *Journal of Hydrology*, 299, 4-14.

Chen, H., Guo, S., Xu, C., Singh, V.P. 2007 Historical temporal trends of hydro-climatic variables and runoff response to climate variability and their relevance in water resource management in the Hanjiang basin. *Journal of Hydrology*, 344, 171-184.

Chesson, P.; Gebauer, R.L.E.; Schwinning, S.; Huntly, N.; Wiegand, K.; Ernest, M.S.K.; Sher, A.; Novoplansky, A.; Weltzin, J.F. 2004. Resource pulses, species interactions, and diversity maintenance in arid and semi-arid environments. *Oecologia*, 141, 236–253.

Chiang, S.-M.; Tsay, T.K.; Nix, S.J.; 2002. Hydrologic regionalization of watersheds. I: methodology development. *Journal of Water Resources Planning and Management*, 128, 3-11.

Chigerwe, J.; Manjengwa, N.; van der Zaag, P.; Zhakata, W.; Rockström, J. 2004. Low head drip irrigation kits and treadle pumps for smallholder farmers in Zimbabwe: a technical evaluation based on laboratory tests. *Physics and Chemistry of the Earth*, 29, 1049-1059.

Chilton, P.J.; Foster, S.S.D. 1995 Hydrogeological characterisation and water-supply potential of basement aquifers in tropical Africa. *Hydrogeology Journal*, 3, 36-49

Chimomwa, M. and Nugent, C. 1993. *A fisheries GIS for Zimbabwe: an initial analysis of the numbers, distribution and size of Zimbabwe's small dams*. Rep. FAO/UNDP ZIM/88/021. Fisheries and Aquaculture Department, Food and Agriculture Organisation of the United Nations. URL: http://www.fao.org/docrep/field/003/AB969E/AB969E00.htm.

Chinoda, G.; Matura, N.; Moyce, W.; Owen, R. 2009. Baseline Report on the Geology of the Limpopo Basin Area, a contribution to the Challenge Program on Water and Food Project 17 "Integrated Water Resource Management for Improved Rural Livelihoods: Managing risk, mitigating drought and improving water productivity in the water scarce Limpopo Basin". *WaterNet Working Paper* 7. WaterNet, Harare. http://www.waternetonline.ihe.nl/workingpapers/WP7 Limpopo Geology.pdf

Christensen, J.H.; Hewitson, B.; Busuioc, A.; Chen, A.; Gao, X.; Held, I.; Jones, R.; Kolli, RK.; Kwon, W.-T.; Laprise, R.; Magaña Rueda, V.; Mearns, L.; Menéndez, C.G.; Räisänen, J.; Rinke, A.; Sarr, A.; Whetton, P. 2007. Regional Climate Projections. In: *Climate Change 2007: The Physical Science Basis. Contribution of Working Group I to the Fourth Assessment Report of the Intergovernmental Panel on Climate Change*, eds.; Solomon, S.D.; Qin, M.; Manning, Z.; Chen, M.; Marquis, K.B.; Averyt, M.; Tignor, A.; Miller, H.L.; Cambridge University Press, Cambridge. pp847-940.

Clarke, D. 1998. *CropWat for Windows: a User Guide*. Food and Agriculture Organization of the United Nations: Rome.

Cobbing, J.E.; Hobbs, P.J.; Meyer, R.; Davies, J. 2008. A critical overview of transboundary aquifers shared by South Africa. *Hydrogeology Journal* 16: 1207-1214.

Cohen, J. 1988. *Statistical Power Analysis for the Behavioural Sciences Analysis*. (2nd edn.). Lawrence Erlbaum, Philadelphia. 567 pp.

Commission for Africa, 2005. *Our Common Interest: The Report of the Africa Commission*. Commission for Africa, London. 462pp

Cooper, P. J. M; Dimes, J; Rao, K. P. C; Shapiro, B; Shiferaw, B; Twomlow, S. 2008. Coping better with current climatic variability in the rain-fed farming systems of sub-Saharan Africa: An essential first step in adapting to future climate change? *Agriculture, Ecosystems & Environment*, 126, 24-35.

Cox, K.G.; Johnson, R.L.; Monkman, L.J.; Stillman, C.J.; Vail, J.R.; Wood, D.N. 1965. The geology of the Nuanetsi igneous province. *Philosophical Transactions of the Royal Society London*, A257, 71-218

CPWF (Challenge Program on Water and Food). 2006. *Integrated Database Information System: Limpopo Basin Kit*. Challenge Program on Water and Food, Colombo.

Cullmann, J.; Wriedt, G. 2008. Joint application of event-based calibration and dynamic identifiability analysis in rainfall–runoff modelling: implications for model parametrisation. *Journal of Hydroinformatics,* 10, 301-316.

Dahlin, T; Owen, R.J.S; 1998. Geophysical investigation of alluvial aquifers in Zimbabwe. In: *Proceedings of the 4th Meeting Environmental and Engineering Geophysics, Barcelona, Spain, 14-17 September 1998*, p 151-154.

Dang, Y.P.; Pringle, M.J.; Schmidt, M.; Dalal, R.C.; Apan, A. 2011. Identifying the spatial variability of soil constraints using multi-year remote sensing. *Field Crops Research*, 123, 248-258.

Davis, S.N. 1969 Porosity and permeabiliy of natural materials. In: *Flow through Porous Media*, ed. De Wiest, R.J.M.; Academic Press, New York, pp54-89

De Groen, M.M.; Savenije, H.H.G. 2006. A monthly interception equation based on the statistical characteristics of daily rainfall, *Water Resources Research*, 42, W12417, doi:10.1029/2006WR005013.

De Fraiture, C.; Wichelns, D. 2010. Satisfying future water demands for agriculture. *Agricultural Water Management* 97, 502–511.

De Hamer, W.; Love, D.; Booij, M.J.; Hoekstra, A. 2007. A simple rainfall-runoff model for an ungauged catchment using the water balance of a reservoir for calibration. In: Abstract volume, *8th WaterNet/WARFSA/GWP-SA Symposium*, Livingstone, Zambia, November 2007, p14. http://doc.utwente.nl/61523/1/rainfall-runoff.pdf

De Hamer, W.; Love, D.; Owen, R.J.S.; Booij, M.J.; Hoekstra, A. 2008. Potential water supply of a small reservoir and alluvial aquifer system in southern Zimbabwe. *Physics and Chemistry of the Earth*, 33, 633-639. **[Annex 1.6 of this thesis]**

De Leon, C.; Douthwaite, B.; Alvarez, S. 2009. Most significant change stories from the Challenge Program on Water and Food. *CPWF Working Paper 03*, The CGIAR Challenge Program on Water and Food, Colombo, Sri Lanka, 93pp.

De Wit, M.; Stankiewicz, J. 2006. Changes in surface water supply across Africa with predicted climate change. *Science*, 311, 1917-1921.

Desanker, P.; Magadza, C. 2001. Africa. In: *Climate Change 2001: The Scientific Basis: Contribution of Working Group I to the Third Assessment Report of the Intergovernmental Panel on Climate Change (IPCC)*, eds., Houghton, J.T.; Ding, Y.; Griggs, M.; Noguer, M.; van der Linden, P.J.; Dai, X.; Maskell, K.; Johnson, C.A., Cambridge University Press, Cambridge, UK, pp487-531.

Dhliwayo, C. 2006. An on-farm comparison of conservation agriculture practices and conventional farmer practices on soil hydrology and maize yield. MSc dissertation (unpublished), Integrated Water Resources Management Programme, University of Zimbabwe.

Didszun, J.; Uhlenbrook, S. 2008. Scaling of dominant runoff generation processes: nested catchments approach using multiple tracers. *Water Resources Research*, 44, W02410, doi:10.1029/2006WR005242.

Dimes, J.; Cooper, P.; Rao, K.P.C. 2009. Climate change impact on crop productivity in the semi-arid tropics of Zimbabwe in the 21st century. In: *Proceedings of the Workshop on Increasing the Productivity and Sustainability* of Rainfed Cropping Systems of Poor, Smallholder Farmers, Tamale, Ghana, 22-25 September 2008, ed., Humphreys, L., The CGIAR Challenge Program on Water and Food, Colombo. Pp189-198.

Dlamini, V.G. 2008. Promoting multiple water use services based on integrated water resources management: the planning and implementation. In: Water Distribution Systems Analysis 2008, ed. Van Zyl,K.; American Society of Civil Engineers ASCE Online, 1-10.

Dong, C.; Schoups, G.; van de Giesen, N. in press. Scenario development for water resource planning and management: A review. Technological Forecasting and Social Change. doi: 10.1016/j.techfore.2012.09.015.

Douville, H.; Chauvin, F.; Broqua, H. 2001. Influence of Soil Moisture on the Asian and African Monsoons. Part I: Mean Monsoon and Daily Precipitation. Journal of Climate, 14, 2381-2403.

DRSS (Department of Research and Specialist Services). 1979. Provisional Soil Map of Zimbabwe Rhodesia, 2nd Edition, DRSS, Harare.

Du Toit, A.S; Prinsloo, M.A. 2001. El Niño-southern oscillation effects on maize production in South Africa: A preliminary study. In: Impact of El Nino and Climate Variability on Agriculture. ASA Special Publication 63, 77-86.

Dubrueil, P.L. 1985. Review of field observations of runoff generation in the tropics. Journal of Hydrology 80, 237-264.

Ekström, K.; Prenning, C.; Dladla, Z. 1996. Geophysical investigations of alluvial aquifers in Zimbabwe. MSc thesis (unpublished), Department of Geotechnology, Lund University, Lund, Sweden.

Engeland, K.; Hisdal, H.; Beldring, S. Predicting low flows in ungauged catchments. In: Climate Variability and Change: Hydrological Impacts, IAHS Publication 308, 163-168.

Engelbrecht, F.A.; Landman, W.A.; Engelbrecht, C.J.; Landman, S.; Bopape, M.M.; Roux, B.; McGregor, J.L.; Thatcher, M. 2011. Multi-scale climate modelling over Southern Africa using a variable-resolution global model. Water SA, 37, 647-658. DOI: 10.4314/wsa.v37i5.2.

Falkenmark, M.; Rockström, J. 2004. Balancing Water for Man and Nature: The new approach in Ecohydrology. EarthScan, London.

Falkenmark, M.; Finlayson, C.M.;Gordon, L.J. 2007. Agriculture, water and ecosystems: avoiding the costs of going too far. Water for Food, Water for Life: A Comprehensive Assessment of Water Management in Agriculture, ed., Molden, D., Earthscan, London and International Water Management Institute, Colombo.

FAO (Food and Agricultural Organisation of the United Nations), 1988. FAO-UNESCO Soil Map of the World. Revised Legend. F.A.O. World Soil Resources Report No. 60. Food and Agricultural Organisation of the United Nations, Rome.

FAO (Food and Agricultural Organisation of the United Nations), 2004. The State of Food Insecurity in the World: Monitoring progress towards the World Food Summit and Millennium Development Goals. Food and Agricultural Organisation of the United Nations, Rome.

FAO (Food and Agricultural Organisation of the United Nations), 2012. Post Harvest Survey: 2011/12 Further analysis. Agriculture Coordination Working Group, Harare. http://www.acwg.co.zw

FAO (Food and Agricultural Organisation of the United Nations), 2013a. Crop Water Information: Alfalfa. http://www.fao.org/nr/water/cropinfo_alfalfa.html

FAO (Food and Agricultural Organisation of the United Nations), 2013b. Crop Water Information: Maize. http://www.fao.org/nr/water/cropinfo_maize.html

FAO (Food and Agricultural Organisation of the United Nations), 2013c. *Crop Water Information: Millet.* http://www.fao.org/nr/water/cropinfo_millet.html

FAO (Food and Agricultural Organisation of the United Nations), 2013d. *Crop Water Information: Sorghum.* http://www.fao.org/nr/water/cropinfo_sorghum.html

Farmer, D.; Sivapalan, M.; Jothitjangkoon, C. 2003. Climate, soil and vegetation controls upon the variability of water balance in temeperate and semiarid landscapes: downward approach to water balance analysis. *Water Resources Research*, 39, WR000328, doi:10.1029/2001WR000328.

Farquharson, F.A.K.; Bullock, A. 1992. The hydrology of basement complex regions of Africa with particular reference to southern Africa. *Geological Society of London Special Publications*, 66, 59-76.

Faulkner, J.W.; Steenhuis, T.; van de Giesen, N.; Andreini, M.; Liebe, J. 2008. Water use and productivity of two small reservoir irrigation schemes in Ghana's Upper East Region, *Irrigation and Drainage*, 57, 151–163.

Faurés, J.-M.; Svendsen, M.; Turral, H. 2007. Reinventing irrigation. *Water for Food, Water for Life: A Comprehensive Assessment of Water Management in Agriculture*, ed., Molden, D., Earthscan, London and International Water Management Institute, Colombo, pp353-394.

Fenicia, F.; Savenije, H.H.G.; Matgen, P.; Pfister, L. 2008. Understanding catchment behavior through stepwise model concept improvement. *Water Resources Research*, 44, WR005563, doi:10.1029/2006WR005563.

Fewsnet, 2012. Zimbabwe Food Security Outlook November 2012 to March 2013. Famine Early Warning Systems Network, www.fews.net

Fung, F.; Lopez, A.; New, M.; 2011. Water availability in +2°C and +4°C worlds. *Philosophical Transactions of the Royal Society A*, 369, 99-116.

Ganoulis, J. 2004. Integrated Risk Analysis for Sustainable Water Resources Management. In: *Comparative Risk Assessment and Environmental Decision Making*, eds. Linkov, I; Ramadan, A.B.D.; NATO Science Series, Kluwer Academic Publishers, Dordrecht. pp. 275-286.

Gebremedhin, B.; Swinton, S. M. 2003. Investment in soil conservation in northern Ethiopia: the role of land tenure security and public programs. *Agricultural Economics*, **29**, 69-84.

Gilbert, R.O. 1987. *Statistical Methods for Environmental Pollution Monitoring.* Wiley, New York, 320 pp.

Giller, K. E; Witter, E; Corbeels, M; Tittonell, P. 2009. Conservation agriculture and smallholder farming in Africa: The heretics' view. *Field Crops Research*, 114, 23-34.

Gómez-Plaza, A.; Martínez-Mena, M.; Castillo, V.M.; 2001. Factors regulating spatial distribution of soil water content in small semiarid catchments. *Journal of Hydrology* ,253, 211-226.

González-Hidalgo, J.C.; de Luis, M.; Raventós, J.; Sánchez, J.R. 2001. Spatial distribution of seasonal rainfall trends in a western Mediterranean area. *International Journal of Climatology*, 21, 843-860.

Gordon, N.D.; Finlayson, B.; McMahon, T.; 2004. *Stream Hydrology.* 2nd edition. John Wiley, New Jersey. 444p.

Görgens, A.H.M.; Boroto, R.A. 1997. Limpopo River: flow balance anomalies, surprises and implications for integrated water resources management. In: *Proceedings of the 8th South African National Hydrology Symposium*, Pretoria, South Africa.

Gumbo, B. 2004. The status of water demand management in selected cities of southern Africa. *Physics and Chemistry of the Earth*, 29, 1225-1231.

Guzman, J.A.; Chu, M.L. 2004. *SPELL-stat v. 1.5.1.10 B*. Grupo en Prediccion y Modelamiento Hidroclimatico, Universidad Industrial de Santander, Colombia.

Gupta, S.C. 2002. On-farm technology transfer: experience of ICRISAT-Nigeria. In: *Improving income and food supply in the Sahel – On-farm testing of sorghum and pearl millet technologies: summary proceedings of the Stakeholders' Workshop to Plan and Implement the IFAD Project, 24-26 February 1999, ICRISAT, Sadoré, Niger*. International Crops Research Institute for the Semi-Arid Tropics, Bamako, Mali and International Fund for Agricultural Development, Rome, Italy. http://oar.icrisat.org/1112/1/RA_00385.pdf

Hanjra, M.A.; Gichuki, F. 2008. Investments in agricultural water management for poverty reduction in Africa: Case studies of Limpopo, Nile, and Volta river basins. *Natural Resources Forum*, 32, 185–202.

Harbaugh, A.W. 2005. MODFLOW-2005, the US Geological Survey modular ground-water module - the Ground-water flow process. *Techniques of Water-Resources Investigations of the United States Geological Survey Book 6, Modeling Techniques, Section A, Ground Water US Geological Survey*, Washington DC, 253p

Harrington, L.W.; Douthwaite, B.; de Leon, C.; Woolley, J. 2008. Stories from the field: a most significant change synthesis. *CPWF Working Paper 02*, The CGIAR Challenge Program on Water and Food, Colombo, Sri Lanka, 43pp.

Hearn, P.Jnr.; Hare, T.; Schruben, P.; Sherill, D.; LaMar, C.; Tsushima, P. 2001. *Global GIS Database: Digital Atlas of Africa*. United States Geological Survey, Washington D.C.

Heinrich, G. 2001. Improving productivity and incomes for small-scale farmers in the semi-arid areas of Zimbabwe: on-farm participatory research in Gwanda. In: *Improving soil management options for women farmers in Malawi and Zimbabwe: proceedings of a Collaborators' Workshop on the DFID-supported Project "Will Women Farmers Invest in Improving their Soil Fertility Management? Participatory Experimentation in a Risky Environment"*, eds., Twomlow, S.J; , Ncube, B., ICRISAT Bulawayo, Zimbabwe. http://oar.icrisat.org/334/1/CO_0012.pdf

Herbert, R. 1998. Water from sand rivers in Botswana. *Quarterly Journal of Engineering Geology and Hydrogeology*, 31, 81-83.

Heuvelmans, G.; Muys, B.; Feyen, J.; 2004. Evaluation of hydrological parameter transferability for simulating the impact of land use on catchment hydrology. *Physics and Chemistry of the Earth*, 29, 739-747.

Hiernaux, P; Mougin, E; Diarra, L; Soumaguel, N; Lavenu, F; Tracol, Y; Diawara, M. 2009. Sahelian rangeland response to changes in rainfall over two decades in the Gourma region, Mali. *Journal of Hydrology*, 375, 114-127.

Hirsch, R.M.; Slack, J.R.; Smith, R.A. 1982. Techniques of trend analysis for monthly water quality data, *Water Resources Research*, 18, 107-121.

Hoffmann, S.; Csitári, G.; Hegedüs, L. 2002. The humus content and soil biological properties as a function of organic and mineral fertilization. *Archives of Agronomy and Soil Science*, 48, 141-146.

Hoko, Z.; 2005. An assessment of drinking ground water quality in rural districts in Zimbabwe: the case of Gokwe South, Nkayi, Lupane, and Mwenezi Districts. *Physics and Chemistry of the Earth*, 30, 859-866.

Homann, S; Rooyen, A.V. 2008. Unexploited Agricultural Growth: The case of crop-livestock production systems in Zimbabwe. In: *Second International*

Conference of the African Association of Agricultural Economists (AAAE), August 20-22, 2007, Accra, Ghana.

Hope, P.K.; Drosdowsky, W.; Nicholls, N. 2006. Shifts in the synoptic systems influencing southwest Western Australia. *Climate Dynamics*, 26, 751-764.

Hopmans, J.W.; Parlange, J.-V.; Assouline, S. 2007. Infiltration. In: *The Handbook of Groundwater Engineering*, ed. Delleur, J.W. ; Taylor and Francis, New York, pp7-1–7-18.

Houston, J. 1988. Rainfall-runoff-recharge relationships in the basement rocks of Zimbabwe. In: *Estimation of Natural Groundwater Recharge*, ed. Simmers, I .; Reidel, Dordrecht, pp349-565

Hughes D.A. 1995. Monthly rainfall-runoff models applied to arid and semiarid catchments for water resource estimation purposes. *Hydrological Sciences*, **40**, 751-769.

Hughes, D.A. 2005. Hydrological issues associated with the determination of environmental water requirements of ephemeral rivers. *River Research and Applications* 8: 899-908. DOI: 10.1002/rra.857.

Hughes, D.A. 2006. Comparison of satellite rainfall data with observations from gauging station networks. *Journal of Hydrology* 327, 399–410

Hughes, D.A.; Kapangaziwiri, E.; Baker, K. 2010. Initial evaluation of a simple coupled surface and ground water hydrological model to assess sustainable ground water abstractions at the regional scale. *Hydrology Research*, 41, 1-12.

Hulme, M.; Doherty, R.M.; Ngara, T.; New, M.G.; Lister, D. 2001. African climate change: 1900–2100. *Climate Research*, 17, 145-168.

Hundecha, Y.; Bárdossy, A. 2005. Trends in daily precipitation and temperature extremes across western Germany in the second half of the 20th century. *International Journal of Climatology*, 25, 1189-1202.

Jaspers, F.G.W. 2003. Institutional arrangements for integrated river basin management. *Water Policy*, 5, 77-90.

Jewsbury, J.M.; Imevbore, A.M.A. 1988. Small dam health studies. *Parasitology Today*, 4, 57-59.

Jewitt, G.P.W. 1992. *Process studies for simulation modelling of forest hydrological impacts*. M.Sc. dissertation (unpublished), University of Natal, Pietermaritzburg, Department of Agricultural Engineering, 145p.

Joel, A.; Messing, I.; Seguel, O.; Casanova, M. 2002. Measurement of surface water runoff from plots of two different sizes. *Hydrological Processes,* 16, 1467-1478.

Johnson, A. I. 1967. Specific yield- compilation of specific yields for various materials. *United States Geological Survey Water Supply Paper, 1662-D*, 74p

Johst, M.; Uhlenbrook, S.; Tilch, N.; Zillgens, B.; Didszun, J.; Kirnbauer, R.; 2008. An attempt of process-oriented rainfall-runoff modelling using multiple-response data in an alpine catchment, Loehnersbach, Austria. *Hydrology Research* 39, 1-16.

Juízo, D and Líden, R. 2008. Modeling for transboundary water resources planning and allocation. *Hydrology and Earth System Sciences Discussions* 5, 475-509.

Jury, M.R.; Mpeta, E.J. 2005. The annual cycle of African climate and its variability. *Water SA*, 31, 1-8.

Kabel, T.C.; 1984. *An Assessment of Surface Water Resources of Zimbabwe and Guidelines for Development Planning*. Government of Zimbabwe, Harare, 16p.

Kapangaziwiri, E.; Hughes, D.A.; 2008. Towards revised physically-based parameter estimation methods for the Pitman monthly rainfall-runoff model. *Water SA*, 32, 183-191.

Kassam, A; Friedrich, T; Shaxson, F; Pretty, J. 2009. The spread of conservation agriculture: Justification, sustainability and uptake. *International Journal of Agricultural Sustainability*, 7, 292-320.

Kaumbotho, P.G; Mwenya, E. 2000. Preserving the environment through conservation tillage with animal traction. In: *Proceedings of the Workshop of the Animal Traction Network for Eastern and Southern Africa ATNESA*, eds., Kaumbotho, P.G; Pearson, R.A; Simalenga, T.A., Mpumalanga, South Africa, 20-24 September 1999.

Khosa, S.; Love, D.; Mul., M. 2008. Evaluation of the effects of different water demand scenarios on downstream water availability: The case of Thuli river basin. In: *Abstract Volume, 9th WaterNet/WARFSA/GWP-SA Symposium*, Johannesburg, South Africa, October 2008, p78. http://www.waternetonline.ihe.nl/challengeprogram/P49%20Khosa%20Water%20demand.pdf **[Annex 1.12 of this thesis]**

Kileshye Onema, J.-M; Van Rooyen, A. 2007. Land use dynamics in a small watershed of the semi-arid Zimbabwe. *American Geophysical Union, Fall Meeting 2007*, abstract #B41B-0460

Kileshye Onema, J.-M.; Mazvimavi, D.; Love, D.; Mul, M. 2006. Effects of dams on river flows of Insiza River, Zimbabwe. *Physics and Chemistry of the Earth*, 31, 870-875. **[Annex 1.2 of this thesis]**

King, J.; Louw, D. 1998. Instream flow assessments for regulated rivers in South Africa using the Building Block Methodology. *Aquatic Ecology Health Management*, 1, 109-124.

King, J.; Brown, C.; Sabet, H. 2003. A scenario-based approach to environmental flow assessments for rivers. *River Research and Applications* 19, 619-639.

Komakech, H.C; Van der Zaag, P. Mul, M.L; Mwakalukwa, T.A; Kemerink, J.S. 2012. Formalization of water allocation systems and impacts on local practices in the Hingilili sub-catchment, Tanzania. *International Journal of River Basin Management*, 10, 213-227

Kondolf, G.M.; Swanson, M.L. 1993. Channel adjustments to reservoir construction and gravel extraction along Stony Creek, California. *Environmental Geology*, 21, 256-269.

Koren, V.; Reed, S.; Smith, M.; Zhang, Z.; Seo, D.J.; 2004. Hydrology laboratory research modeling system (HL-RMS) of the US national weather service. *Journal of Hydrology*, 291, 297-318.

KwaZulu-Natal Department of Agriculture and Environmental Affairs. 2008. Beef Production: the Basics. *Agricultural Production Guidelines*. Pietermaritzburg, South Africa.

Lal. R. 1997. Long-term tillage and maize monoculture effects on a tropical Alfisol in western Nigeria. I. Crop yield and soil physical properties. *Soil and Tillage Research*, 42, 145-160.

Lange, J. 2005. Dynamics of transmission losses in a large arid stream channel. *Journal of Hydrology*, 306, 112-126.

Lange, J.; Leinbundgut, C. 2003. Surface runoff and sediment dynamics in arid and semi-arid regions. In: *International Contributions to Hydrogeology 238: Understanding Water in a Dry Environment: Hydrological processes in arid and semi-arid zones*, ed., Simmers, I., Balkema, Rotterdam. pp114-150.

Larsen, F.; Owen, R.; Dahlin, T.; Mangeya, P.; Barmen, G. 2002. A preliminary analysis of the groundwater recharge to the Karoo formations, mid-Zambezi basin, Zimbabwe. *Physics and Chemistry of the Earth*, 27, 765–772.

Lasage, R.; Aerts, J.; Mutiso, G.-C.M.; de Vries, A. 2007. Potential for community based adaptation to droughts, sand dams in Kitui, Kenya. *Physics and Chemistry of the Earth*, 33, 67-73

Laslett, D.; Davis, G.B. 1998. Using HST3D to model flow and tracer transport in an unconfined sandy aquifer with large drawdown. In: *International Contributions to Hydrogeology 18, Shallow Groundwater Systems*, eds., Dillon, P.; Simmers, I.; Balkema, Rotterdam, pp207-219

Latham, C.J.K.; 2001. Manyame Catchment Council: a review of the reform of the water sector in Zimbabwe. *Physics and Chemistry of the Earth*, 27, 907-918.

Lebel, T.; Delcalux, F.; Le Barbé, L.; Polcher, J. 2000. From GCM scales to hydrological scales: rainfall variability in West Africa. *Stochastic Environmental Research and Risk Assessment*, 14, 275-295.

Li, X.; Su, D.; Yuan, Q. 2007. Ridge-furrow planting of alfalfa (*Medicago sativa* L.) for improved rainwater harvest in rainfed semiarid areas in Northwest China. *Soil and Tillage Research*, 93, 117-125.

Lidén, R.; 2000. Extending the hydrological memory in HBV-96 simulations in arid regions. *Proceedings of the XXI Nordic Hydrological Conference, Uppsala, Sweden*, 2, 367-373.

Lidén, R. and Harlin, J. 2000. Analysis of conceptual rainfall–runoff modelling performance in different climates. *Journal of Hydrology*, 238, 231-247.

Lin, Z.; Radcliffe, D.E.; 2006. Automatic calibration and predictive uncertainty analysis of a semidistributed watershed model. *Vadose Zone Journal* 5,248-260.

Littlewood, I.G.; Jakeman, T.; Croke, B.; Kokkonen, T.; Post, D.; 2002. Unit hydrograph characterisation of flow regimes leading to streamflow estimation in ungauged catchments (regionalization). In: *Predictions in Ungauged Basins: PUB Kick-off (Proceedings of the PUB Kick-off meeting held in Brasilia, 20–22 November 2002), IAHS Publication 309*, pp45-52.

Logan, J.A.; Megretskaia, I.; Nassar, R.; Murray, L.T.; Zhang, L.; Bowman, K.W.; Worden, H.M.; Luo, M. 2008. Effects of the 2006 El Niño on tropospheric composition as revealed by data from the Tropospheric Emission Spectrometer (TES). *Geophysics Research Letters*, 35, L03816, doi:10.1029/2007GL031698.

Longobardi, A.; Villani, P.; 2008. Baseflow index regionalization analysis in a Mediterranean area and data scarcity context: Role of the catchment permeability index. *Journal of Hydrology* 355, Pages 63-75.

Loucks, D. P.; van Beek, E.; 2005. *Water Resources Systems Planning and Management: An Introduction to Methods, Models and Applications*, UNESCO, Paris. 680p.

Love, D and Walsh, KL. 2009. Geological evidence does not support suggestions of mining in the Nyanga upland culture. *Prehistory Society of Zimbabwe Newsletter*, 140, 1-6.

Love, D; Jonker, L; Rockström J; van der Zaag, P; Twomlow, S. 2004. The Challenge of Integrated Water Resource Management for Improved Rural Livelihoods in the Limpopo Basin – an introduction to WaterNet's first network research program. In: Abstract volume, 5th WaterNet-WARFSA Symposium, Windhoek, Namibia, pp106-107.

Love, D.; Twomlow, S.; Mupangwa, W.; van der Zaag, P.; Gumbo, B. 2006a. Implementing the millennium development food security goals - challenges of the southern African context. *Physics and Chemistry of the Earth*, 31, 731-737. **[Chapter 2 of this thesis]**

Love, D.; Moyce, W.; Ravengai, S. 2006b. Livelihood challenges posed by water quality in the Mzingwane and Thuli river catchments, Zimbabwe. 7th

WaterNet/WARFSA/GWP-SA Symposium, Lilongwe, Malawi, November 2006. http://www.waternetonline.ihe.nl/challengeprogram/P17%20Love%20water%20 quality.pdf **[Annex 1.15 of this thesis]**

Love, F.; Madamombe, E.; Marshall, B.; Kaseke, E. 2006c. Estimating Environmental Flow Requirements of the Rusape River, Zimbabwe. *Physics and Chemistry of the Earth*, 31,, 864-869.

Love, D. de Hamer, W.; Owen, R.J.S.; Booij, M.J.; Uhlenbrook, S.; Hoekstra, A.; van der Zaag, P. 2007. Case studies of groundwater – surface water interactions and scale relationships in small alluvial aquifers. In: *Abstract volume, 8th WaterNet/WARFSA/GWP-SA Symposium*, Lusaka, Zambia, November 2007, p21. **[Annex 1.11 of this thesis]**

Love, D.; Love, F.; van der Zaag, P.; Uhlenbrook, S.; Owen, R.J.S. 2008a. Impact of the Zhovhe Dam on the lower Mzingwane River channel. In: *Fighting Poverty Through Sustainable Water Use: Proceedings of the CGIAR Challenge Program on Water and Food 2nd International Forum on Water and Food, Addis Ababa, Ethiopia, November 10 - 14 2008*, IV, eds. Humphreys, E.; Bayot, R.S.; van Brakel, M.; Gichuki, F.; Svendsen, M.; Wester, P.; Huber-Lee, A.; Cook, S.; Douthwaite, B.; Hoanh, C.T.; Johnson, N.; Nguyen-Khoa, S.; Vidal, A.; MacIntyre, I.;. The CGIAR Challenge Program on Water and Food, Colombo, pp46-51 **[Section 7.6 of this thesis]**

Love, D.; Khosa, S.; Mul, M.; Uhlenbrook, S.; van der Zaag, P. 2008b. Modelling upstream-downstream interactions using a spreadsheet-based water balance model: two case studies from the Limpopo basin. In: *Fighting Poverty Through Sustainable Water Use: Proceedings of the CGIAR Challenge Program on Water and Food 2nd International Forum on Water and Food, Addis Ababa, Ethiopia, November 10 - 14 2008*, IV, eds. Humphreys, E.; Bayot, R.S.; van Brakel, M.; Gichuki, F.; Svendsen, M.; Wester, P.; Huber-Lee, A.; Cook, S.; Douthwaite, B.; Hoanh, C.T.; Johnson, N.; Nguyen-Khoa, S.; Vidal, A.; MacIntyre, IV.;. The CGIAR Challenge Program on Water and Food, Colombo, pp15-21 **[Annex 1.8 of this thesis]**

Love, D.; Uhlenbrook, S.; Corzo-Perez, G.; Twomlow, S.; van der Zaag, P. 2010a. Rainfall-interception-evaporation-runoff relationships in a semi-arid catchment, northern Limpopo Basin, Zimbabwe. *Hydrological Sciences Journal*, **55**, 687-703. **[Chapter 4 of this thesis]**

Love, D.; Uhlenbrook, S.; Twomlow, S.; van der Zaag, P. 2010b. Changing rainfall and discharge patterns in the northern Limpopo Basin, Zimbabwe. *Water SA*, 36, 335-350. **[Chapter 3 of this thesis]**

Love, D.; van der Zaag, P.; Uhlenbrook, S.; Owen, R. 2010c. A water balance modelling approach to optimising the use of water resources in ephemeral sand rivers. *River Research and Applications*, **26**, 908-925. DOI: 10.1002/rra.1408 **[Chapter 7 of this thesis]**

Love, D.; Uhlenbrook, S.; van der Zaag, P. 2011. Regionalising a meso-catchment scale conceptual model for river basin management in the semi-arid environment. *Physics and Chemistry of the Earth*, 36, 747-760. **[Chapter 5 of this thesis]**

Love, D.; Owen, R.J.S.; Uhlenbrook, S.; van der Zaag, P. *submitted*. Targeting the under-valued resource: an evaluation of the water supply potential of small sand rivers in the northern Limpopo Basin. Submitted to *Water Resources Management*. **[Chapter 6 of this thesis]**

Lupankwa, K.; Love, D.; Mapani, B.S.; Mseka, S. 2004. Impact of a base metal slimes dam on water systems, Madziwa Mine, Zimbabwe. *Physics and Chemistry of the Earth*, 29, 1145-1151.

Magombeyi, M. S.; Taigbenu, A. E. 2008. Crop yield risk analysis and mitigation of smallholder farmers at quaternary catchment level: Case study of B72A in Olifants river basin, South Africa. *Physics and Chemistry of the Earth*, 338, 744-756.

Mahe, G; Paturel, J.-E; Servat, E; Conway, D; Dezetter, A. 2005. The impact of land use change on soil water holding capacity and river flow modelling in the Nakambe River, Burkina Faso. *Journal of Hydrology*, 300, 33-43.

Maisiri, N.; Rockström, J.; Senzanje, A.; Twomlow, S. 2005. An on-farm evaluation of the effects of low cost drip irrigation on water and crop productivity, compared to conventional surface irrigation system. *Physics and Chemistry of the Earth*, 30, 783-791.

Makarau, A.; Jury, M.R. 1997. Predictability of Zimbabwe summer rainfall. International *Journal of Climatology*, 17, 1421-1432.

Mandiziba, W. 2008. *The effects of suspended solids on the yield of sand abstraction systems*. Honours Project report unpublished, Department of Civil and Water Engineering, the National University of Science and Technology, Bulawayo, Zimbabwe, 56p

Mansell, M.G.; Hussey, S.W. 2005. An investigation of flows and losses within the alluvial sands of ephemeral rivers in Zimbabwe. *Journal of Hydrology*, 314, 192-203

Masih I; Uhlenbrook S; Maskey S; Ahmad M.D; 2010: Regionalization of a conceptual rainfall–runoff model based on similarity of the flow duration curve: A case study from the semi-arid Karkheh basin, Iran. *Journal of Hydrology*, 391, 188–201.

Manyanga, M. 2006. *Resilient Landscapes: socio-environmental dynamics in the Shashi-Limpopo Basin, southern Zimbabwe c. AD 800 to the present*. Ph.D. thesis (unpublished), Uppsala University Department of Archaeology and Ancient History, 301pp.

Manzungu, E.; Mabiza, C. 2004. Status of water governance in urban areas in Zimbabwe: some preliminary observations from the City of Harare. *Physics and Chemistry of the Earth*, 29, 1167-1172.

Mapedza, E.; Wright, J.; Fawcett, R. 2003. An investigation of land cover change in Mafungautsi Forest, Zimbabwe, using GIS and participatory mapping. *Applied Geography*, 23, 1-21.

Marcé, R.; Ruiz, C.E.; Armengol, J.; 2008. Using spatially distributed parameters and multi-response objective functions to solve parameterization of complex applications of semi-distributed hydrological models. *Water Resources Research* 44, W02436, doi:10.1029/2006WR005785.

Martinez-Mena, M.; Castillo, V.; Albaladejo, J. 1998. Hydrological and erosional response to natural rainfall in a semi-arid area of south-east Spain. *Hydrological Processes*, 15, 557-571.

Masih, I.; Uhlenbrook, S.; Maskey, S.; Ahmad, M.D.; Islam, M.D.A.; 2008. Estimating ungauged stream flows based on model regionalization – Examples from the mountainous, semi-arid Karkheh river basin, Iran. In. *HydroPredict 2008 Conference on Predictions for Hydrology, Ecology, and Water Resources Management: Using Data and Models to Benefit Society 15-18 September 2008*, eds. Brhuthans, J.; Kovar, K.; Hrkal, Z.; Prague, Czech Republic. 7-10 pp

Mason, S.J.; Jury, M.R.; Tyson, P.D. 1997. Climatic variability and change over southern Africa : a reflection on underlying processes. *Progress in Physical Geography*, 21, 23-50.

Masvopo, T.; Love, D.; Makurira, H. 2008. Evaluation of the groundwater potential of the Malala Alluvial Aquifer, Lower Mzingwane River, Zimbabwe. In, *Abstract Volume, 9th WaterNet/WARFSA/GWP-SA Symposium, Johannesburg, South Africa, October 2008*, p7 **[Annex 1.13 of this thesis]**

Matter, J.N.; Waber, H.N.; Loew, S.; Matter, A. 2005. Recharge areas and geochemical evolution of groundwater in an alluvial system in the Sultanate of Oman. *Hydrogeology Journal*, 14, 203-224.

Mazvimavi, D. 2003. *Estimation of Flow Characteristics of Ungauged Catchments: Case study in Zimbabwe*. PhD dissertation, Wageningen University and International Institute for Geo-Information Science and Earth Observation, Delft. 188pp.

Mazvimavi, D. 2010. Investigating changes over time of annual rainfall in Zimbabwe. *Hydrology and Earth System Sciences*, 14, 2671-2679.

Mazvimavi, K; Twomlow, S. 2009. Socioeconomic and institutional factors influencing adoption of conservation farming by vulnerable households in Zimbabwe. *Agricultural Systems*, 101, 20-29.

Mazvimavi, D.; Meijerink, A.M.J.; Savenije, H.H.G.; Stein, A.; 2005. Prediction of flow characteristics using multiple regression and neural networks: a case study in Zimbabwe. *Physics and Chemistry of the Earth* 30, 639-647.

MCC (Mzingwane Catchment Council), 2009. *Mzingwane River System Outline Plan*, Mzingwane Catchment Council, Bulawayo, Zimbabwe, 88pp.

McCartney, M. 2000. The water budget of a headwater catchment containing a dambo. *Physics and Chemistry of the Earth, Part B*, 25, 611-616.

McCartney, M.; Butterworth, J.; Moriarty, P.; Owen, R.J.S. 1998. Comparison of the hydrology of two contrasting headwater catchments in Zimbabwe. In: *Hydrology, Water Resources and Ecology in Headwaters - Proceedings of the Headwaters'98 conference held in Merano, Italy 20-23 April 1998*, IAHS publication. No. 248, eds. Kovar, K.; Tappeiner, U.; Peters, N.E.; Craig, R.G.; pp515-522.

Milly, P.C.D.; Dunne, K.A.; Vecchia, A.V. 2005. Global pattern of trends in streamflow and water availability in a changing climate. *Nature*, 438, 347-350.

Milly, P.C.D.; Betancourt, J.; Falkenmark, M.; Hirsch, R.M.; Kundzewicz, Z.W.; Lettenmaier, D.S.; Stouffer, R.J. 2008. Stationarity is dead: whither water management? *Science*, 319, 573-574.

Milton, S.J; Dean, W.R.J; Richardson, D.M. 2003. Economic incentives for restoring natural capital in southern African rangelands. *Frontiers in Ecology and Environment*, 1, 247-254.

Ministry of Local Government, Rural and Urban Development. 1996. *Matabeleland - Zambezi Water supply feasibility study*, 1, Government of Zimbabwe.

Minshull, J.L. 2008. Dry season fish survival in isolated pools and within sand-beds in the Mzingwane River, Zimbabwe. *African Journal of Aquatic Science*, **33**, 95-98.

Misselhorn, A.A. 2005. What drives food insecurity in southern Africa? a meta-analysis of household economy studies. *Global Environmental Change*, 15, 33-43.

Motsi, K.E; Chuma, E; Mukamuri, B.B. 2004. Rainwater harvesting for sustainable agriculture in communal lands of Zimbabwe. *Physics and Chemistry of the Earth*, 29, 1069-1073.

Moyce, W.; Mangeya, P.; Owen, R.; Love, D. 2006. Alluvial aquifers in the Mzingwane Catchment: their distribution, properties, current usage and potential expansion. *Physics and Chemistry of the Earth*, 31, 988-994. **[Annex 1.1 of this thesis]**

Moyce, W; Meck, M; Owen, R; Love, D. 2010. Influence of basalt weathering on shallow groundwater quality in semi-arid Cawoods-Mazunga, Zimbabwe: petrographic study. In: 3rd IASTED African Conference on Water Resource Management, AfricaWRM2010, Gaborone, Botswana, September 2010. www.actapress.com/Abstract.aspx?paperId=41468

Moyo, S. 2004. Land allocation, beneficiaries and agrarian structure. Paper presented at *National Stakeholders' Dialogue on Land and Agrarian Reform*, Harare, July 2004.

Moyo, B; Madamombe, E; Love, D. 2005. A model for reservoir yield under climate change scenarios for the water-stressed City of Bulawayo, Zimbabwe. In: Abstract Volume, 6th WaterNet-WARFSA-GWP Symposium, Swaziland, November 2005, p38. http://www.bscw.ihe.nl/pub/bscw.cgi/d2606714/MoyoB.pdf **[Annex 1.16 of this thesis]**

Moyo, R.; Love, D.; Mul, M.; Twomlow, S.; Mupangwa, W. 2006. Impact and sustainability of low-head drip irrigation kits, in the semi-arid Gwanda and Beitbridge Districts, Mzingwane Catchment, Limpopo Basin, Zimbabwe. *Physics and Chemistry of the Earth*, 31, 885-892. **[Annex 1.4 of this thesis]**

Mugabe, F.T.; Hodnett, M.; Senzanje, A.; 2007. Comparative hydrological behaviour of two small catchments in semi-arid Zimbabwe. *Journal of Arid Environments*, 69, 599-616.

Mujere, J; Gumbo, S. 2011. Large scale investment projects and land grabs in Zimbabwe: the case of Nuanetsi Ranch Bio-Diesel Project. Paper presented at the International Conference on Global Land Grabbing, 6-8 April 2011.

Mulder, V.L.; De Bruin, S.; Schaepman, M.E.; Mayr, T.R. 2011. The use of remote sensing in soil and terrain mapping—A review. *Geoderma*, 162, 1-19.

Munamati, M; Mhizha, A; Sithole, P. 2005. Cultivating livelihoods: an assessment of water allocation and management practices in small-scale irrigation schemes - case studies in Mzingwane Catchment. In: *Abstract Volume, 6th WaterNet-WARFSA-GWP Symposium, Swaziland, November 2005*, p120.

Mupangwa, W.; Love, D.; Twomlow, S. 2006. Soil-water conservation and other rainwater harvesting strategies in the semi-arid Mzingwane Catchments, Limpopo Basin, Zimbabwe. *Physics and Chemistry of the Earth*, 31, 893-900. **[Annex 1.3 of this thesis]**

Mupangwa, W.; Twomlow, S.; Walker, S.; Hove, L.; 2007. Effect of minimum tillage and mulching on maize (*Zea mays* L.) yield and water content of clayey and sandy soils. *Physics and Chemistry of the Earth*, 32, 1127-1134.

Mupangwa, W; Twomlow, S; Walker, S. 2008. The influence of conservation tillage methods on soil water regimes in semi-arid southern Zimbabwe. *Physics and Chemistry of the Earth*, **33**, 762–767.

Mupangwa, W; Walker, S.; Twomlow, S. 2011. Start, end and dry spells of the growing season in semi-arid southern Zimbabwe. *Journal of Arid Environments*, 75, 1097-1104.

Mupangwa, W; Twomlow, S; Walker, S. 2012. Reduced tillage, mulching and rotational effects on maize (*Zea mays* L.), cowpea (*Vigna unguiculata* (Walp) L.) and sorghum (*Sorghum bicolor* L.(Moench)) yields under semi-arid conditions. *Field Crops Research*, 132, 139-148.

Mupangwa, W; Twomlow, S; Walker, S. 2013. Cumulative effects of reduced tillage and mulching on soil properties under semi-arid conditions. *Journal of Arid Environments*, 91, 1-8.

Mutale, M. 1994. *Assessment of water resources with the help of water quality*. MSc thesis (unpublished), IHE-Delft, the Netherlands, rep. no. HH180.

Mutezo, M.; 2005. Government to rehabilitate irrigation infrastructure. *Interview of the Minister of Water Resources and Infrastructure Development by News hour*, Zimbabwe Broadcasting Holdings, Harare.

Mutezo, M.; 2008. Minister visits Silalatshani. *News hour*, Zimbabwe Broadcasting Holdings, Harare.

Mwakalila, S.; Feyen, J.; Wyseure, G.; 2002. The influence of physical catchment properties on baseflow in semi-arid environments. *Journal of Arid Environments*, 52, 245-258.

Mwenge Kahinda, J.-M.; Rockström, J.; Taigbenu, A.E.; Dimes, J. 2007. Rainwater harvesting to enhance water productivity of rainfed agriculture in the semi-arid Zimbabwe. *Physics and Chemistry of the Earth*, 32, 1068-1073.

Myers, N; (ed.), 1985. *The Gaia Atlas of Planet Management*. Pan Books, London.

Myers, J.L.; Well, A.D. 2002. *Research Design and Statistical Analysis* (2nd edn.). Lawrence Erlbaum, Philadelphia, 760pp.

Nare, L.; Love, D.; Hoko, Z. 2006. Involvement of stakeholders in the water quality monitoring and surveillance system: the case of Mzingwane Catchment. *Physics and Chemistry of the Earth*, 31, 707-712. **[Annex 1.5 of this thesis]**

Nash, J. E; Sutcliffe, J.V. 1970, River flow forecasting through conceptual models part I — A discussion of principles, *Journal of Hydrology*, 10, 282–290.

Ncube, B.; Magombeyi, M.; Munguambe, P.; Mupangwa, W.; Love, D. 2009. Methodologies and case studies for investigating upstream-downstream interactions of rainwater water harvesting in the Limpopo Basin. In: *Proceedings of the Workshop on Increasing the Productivity and Sustainability of Rainfed Cropping Systems of Poor, Smallholder Farmers, Tamale, Ghana, 22-25 September 2008*, ed. Humphreys, L.; The CGIAR Challenge Program on Water and Food, Colombo, 209-221. **[Annex 1.9 of this thesis]**

Ncube, B.; Manzungu, E.; Love, D.; Magombeyi, M.; Gumbo, B.; Lupankwa, K. 2010. The Challenge of Integrated Water Resource Management for Improved Rural Livelihoods: Managing Risk, Mitigating Drought and Improving Water Productivity in the Water Scarce Limpopo Basin. *Challenge Program on Water and Food Project Report* 17, 161p. **[Annex 1.10 of this thesis]**

Ndomba, P.; Mtalo, F.; Killingtveit, A.; 2008. SWAT model application in a data scarce tropical complex catchment in Tanzania. *Physics and Chemistry of the Earth* 33, 626-632.

New, M.; Heiwitson, B.; Stephenson, D.B.; Tsiga, A.; Kruger, A.; Manhique, A.; Gomez, B.; Coelho, C.A.S.; Masisi, D.M.; Kululanga, E.; Mbambalala, E.; Adesina, F.; Saleh, H.; Kanyanga, J.; Adosi, J.; Bulane, L.; Fortunata, L.; Mdoka, M.L.; Lajoie, R. 2006. Evidence of trends in daily climate extremes over southern and west Africa. *Journal of Geophysical Research*, 111, D14102 doi:101029/2005JD006289.

Ngwenya, P.T. 2006. *Effect of soil degradation from grazing pressure on rangeland soil hydrology*. MSc dissertation (unpublished), Integrated Water Resources Management Programme, University of Zimbabwe.

Niadas, I.A.; 2005. Regional flow duration curve estimation in small ungauged catchments using instantaneous flow measurements and a censored data approach. *Journal of Hydrology* 314, 48-66.

Nkomo, D.; van der Zaag, P. 2004. Equitable water allocation in a heavily committed international catchment area: the case of the Komati Catchment. *Physics and Chemistry of the Earth,* 29, 1309–1317.

Nord, M. 1985. *Sand Rivers of Botswana, Results from phase 2 of the Sand Rivers Project.* Unpublished report, Department of Water Affairs, Government of Botswana, Gaborone.

Noto, L.V.; Ivanov, V.Y.; Bras, R.L.; Vivoni, E.R.; 2008. Effects of initialization on response of a fully-distributed hydrologic model. *Journal of Hydrology* 352, 107-125.

Nyabeze, W.R.; 2000. Application of a GIS assisted hydrological drought analysis tool on selected catchments in Zimbabwe. *1st WaterNet/WARFSA Symposium, Maputo, Mozambique, November 2000.*

Nyabeze, W.R.; 2002. Determining parameter reliability levels in a digital information system for monitoring and managing hydrological droughts. *Physics and Chemistry of the Earth* 27, 793-799.

Nyabeze, W.R. 2004. Estimating and interpreting hydrological drought indices using a selected catchment in Zimbabwe. *Physics and Chemistry of the Earth,* 29, 1173-1180.

Nyabeze, W.R.; 2005. Calibrating a distributed model to estimate runoff for ungauged catchments in Zimbabwe. *Physics and Chemistry of the Earth* 30, 625-633.

Nyabeze, W.R.; Love, D. 2007. *CP17 Impact Pathway.* WaterNet, Harare.

Nyamudeza, P; 1999. Agronomic practices for the low rainfall regions of Zimbabwe. In: *Water for agriculture in Zimbabwe,* eds., Manzungu, E; Senzanje, A; van der Zaag, P., University of Zimbabwe Press, Harare.

Olufayo, O.O.; Otieno, F.A.O.; Ochieng, G.M. 2010. Sand water storage systems using coefficient of uniformity as surrogate for optimal design: A laboratory study. *International Journal of the Physical Sciences,* 9, 1227-1230.

Olsthoorn, T.N. 1985. The power of the electronic worksheet: modeling without special programs. *Ground Water* 23, 381-390.

Ofosu, E.A.; van der Zaag, P.; ven de Giesen, N.; Odai, S.N. 2010. Productivity of irrigation technologies in the White Volta Basin. *Physics and Chemistry of the Earth,* 35, 706-716.

Otieno, F.A.O., Olufayo, O.A., Ochieng, G.M. 2011. Sand water storage: Unconventional methods to freshwater augmentation in isolated rural communities of South Africa. *Scientific research and essays,* 4, 1885-1890.

Ott, B.; Uhlenbrook, S.; 2004. Quantifying the impact of land-use changes at the event and seasonal time scale using a process-oriented catchment model. *Hydrology and Earth System Sciences* 8, 62-78.

Oudin, L.; Hervieu, F.; Michel, C.; Perrin, C.; Andréassian, V.; Anctil, F.; Loumagne, C. 2005. Which potential evapotranspiration input for a lumped rainfall–runoff model? Part 2—Towards a simple and efficient potential evapotranspiration model for rainfall–runoff modelling. *Journal of Hydrology,* 303, 290-306.

Owen, R. 1991. *Water Resources for Small-Scale Irrigation from Shallow Alluvial Aquifers in the Communal Lands of Zimbabwe.* MPhil thesis unpublished, Department of Civil Engineering, University of Zimbabwe

Owen, R. 2000. *Conceptual models for the evolution of groundwater flow paths in shallow aquifers in Zimbabwe.* D.Phil. thesis (unpublished), Department of Geology, University of Zimbabwe.

Owen, R.; Dahlin T. 2005. Alluvial aquifers at geological boundaries: geophysical investigations and groundwater resources. In: *Groundwater and Human Development*, eds., Bocanegra, E.; Hernandez M.; Usunoff E., AA Balkema Publishers, Rotterdam. pp233-246.

Owen, R.; Madari, N. .2009 Baseline Report on the Hydrogeology of the Limpopo Basin, country studies from Mozambique, South Africa and Zimbabwe, a contribution to the Challenge Program on Water and Food Project 17 "Integrated Water Resource Management for Improved Rural Livelihoods, Managing risk, mitigating drought and improving water productivity in the water scarce Limpopo Basin". WaterNet Working Paper 12 WaterNet, Harare. http://www.waternetonline.ihe.nl/workingpapers/WP12 Limpopo Hydrogeology.pdf

Owen, R.; Maziti, A.; Dahlin, T. 2007. The relationship between regional stress field, fracture orientation and depth of weathering and implications for groundwater prospecting in crystalline rocks. *Hydrogeology Journal*, 15, 1231-1238.

Palma, H.C.; Bentley, L.R. 2007. A regional-scale groundwater flow model for the Leon-Chinandega aquifer, Nicaragua. *Hydrogeology Journal*, 15, 1457-1472.

Panik, M.; 2005. *Advanced Statistics from an Elementary Point of View*, Elsevier, Amsterdam. 802p.

Parliament of Zimbabwe. 2008. Mwenezi Constituency National Assembly Constituency Information Profile http://www.parlzim.gov.zw/cms/Constituencyinfo/Mwenezi.pdf

Peugeot, C, Cappelaere, B.; Bieux, B.E.; Séguis, L.; Maia, A.; 2003. Hydrologic process simulation of a semiarid, endoreic catchment in Sahelian West Niger. 1. Model-aided data analysis and screening. *Journal of Hydrology* 279, 224-243.

Pettit, A.N. 1979. A non-parametric approach to the change-point problem. *Applied Statistics*, 28, 126-135.

Prundeda, E.B.; Barber, M.E.; Allen, D.M.; Wu, J. 2010. Use of stream response functions to determine impacts of replacing surface-water use with groundwater withdrawals. *Hydrogeology Journal*, 18, 1077-1092

Quillis, R.O.; Hoogmoed, R.M.; Ertsen, M.; Foppen, J.W.; De Vries, A. 2009. Measuring and modelling hydrological processes of sand-storage dams on different spatial scales. *Physics and Chemistry of the Earth*, 34, 289-298.

Raju, N.J.; Reddy, T.V.K.; Munirathnam, P. 2006. Subsurface dams to harvest rainwater - a case study of the Swaranamukhi River basin, southern India. *Hydrogeology Journal*, 14, 526-531.

Ramachandra Rao, A.; Srinivas, V.V.; 2003. Some problems in regionalization of watersheds. In: Water resources systems – water availability and global change (Proceedings of symposium HS02a held during IUGG2003 at Sapporo, July 2003), IAHS Publication 280, pp301-308.

Ravengai, S.; Owen, R.J.S.; Love, D. 2004. Evaluation of seepage and acid generation potential from evaporation ponds, Iron Duke Pyrite Mine, Mazowe Valley, Zimbabwe. *Physics and Chemistry of the Earth*, 29, 1129-1134.

Ravengai, S.; Love, D.; Love, I.; Gratwicke, B.; Mandingaisa, O.; Owen, R. 2005*a*. Impact of Iron Duke Pyrite Mine on water chemistry and aquatic life – Mazowe valley, Zimbabwe. *Water SA*, 31, 219-228.

Ravengai, S.; Love, D.; Mabvira-Meck, M.L.; Musiwa, K.; Moyce, W. 2005*b*. Water quality in an abandoned mining belt, Beatrice, Zimbabwe. *Physics and Chemistry of the Earth*, 30, 826-831.

Richardson, C. J. 2007. How much did droughts matter? Linking rainfall and GDP growth in Zimbabwe. *African Affairs*, 106, 463-478.

Rockström, J; Barron, J; Fox, P. 2003. Water productivity in rain-fed agriculture: challenges and opportunities for smallholder farmers in drought-prone tropical agroecosystems. In: *Water Productivity in Agriculture: Limits and Opportunities for Improvement*, eds., Kijne, J.W; Barker, R.; Molden, D., CAB International, London.

Rodríguez, L.B.; Cello, P.A.; Vionnet, C.A. 2006. Modeling stream-aquifer interactions in a shallow aquifer, Choele-Choel Island, Patagonia, Argentina. *Hydrogeology Journal*, 14, 591-602

Rogers, R.R.; Rogers, K.C.; Munyikwa, D.; Terry, R.C.; Singer, B.S. 2004. Sedimentology and taphonomy of the upper Karoo-equivalent Mpandi Formation in the Tuli Basin of Zimbabwe, with a new ^{40}Ar/^{39}Ar age for the Tuli basalts. *Journal of African Earth Sciences*, 40, 147-161

Rohrbach, D.D. 2001. Zimbabwe baseline: crop management options and investment priorities in Tsholotsho. In: *Improving soil management options for women farmers in Malawi and Zimbabwe: proceedings of a Collaborators' Workshop on the DFID-supported Project "Will Women Farmers Invest in Improving their Soil Fertility Management? Participatory Experimentation in a Risky Environment"*, eds., Twomlow, S.J; , Ncube, B., ICRISAT Bulawayo, Zimbabwe. http://oar.icrisat.org/334/1/CO_0012.pdf

Rollinson, H.; Blenkinsop, T.G. 1995. The magmatic, metamorphic and tectonic evolution of the Northern Marginal Zone of the Limpopo Belt in Zimbabwe. *Journal of the Geological Society*, 152, 65-75

Ryan, J.G; Spencer, D.C. 2001. Future challenges and opportunities for agricultural R&S in the semi-arid tropics. International Crops Research Institute ICRISAT, Patancheru, India.

SADC, 2011. *Climate Change Adaptation in SADC: a Strategy for the Water Sector*. Southern African Development Community, Gaborone. 38pp.

Sakahuni, C; Dzingirai, V; Manzungu, E; Sibanda, T; Ncube, P; Rosen, T. 2012. The relevance and applicability of performance indicators in the Limpopo River Basin in Zimbabwe. *Economics, Management and Financial Markets*, 1-2012, 39-66.

Samakande, I.; Senzanje, A.; Manzungu, E. 2004. Sustainable water management in smallholder irrigation schemes: Understanding the impact of field watyer management on maize productivity on two irrigation schemes in Zimbabwe. *Physics and Chemistry of the Earth*, 29, 1075-1081.

Sanchez, P.A; Swaminathan, M.S. 2005. Hunger in Africa: the link between unhealthy people and unhealthy soils. *Lancet*, 365, 442-444.

Sanders, J. 2002. Input and output markets and the introduction of sorghum-millet technologies. In: *Improving income and food supply in the Sahel – On-farm testing of sorghum and pearl millet technologies: summary proceedings of the Stakeholders' Workshop to Plan and Implement the IFAD Project, 24-26 February 1999, ICRISAT, Sadoré, Niger*. International Crops Research Institute for the Semi-Arid Tropics, Bamako, Mali and International Fund for Agricultural Development, Rome, Italy. http://oar.icrisat.org/1112/1/RA_00385.pdf

Savenije, H.H.G. 1995. Spreadsheets: flexible tools for integrated management of water resources in river basins. In: Modelling and Management of Sustainable Basin-scale Water Resources Systems. *IAHS Publications*, 231, pp. 207–215.

Savenije, H.H.G; 1998. "How do we feed a growing world population in a situation of water scarcity?" 8th Stockholm Symposium "Water - The Key to Socio-economic Development and Quality of Life", SIWI, Stockholm, pp.49-58.

Savenije, H.H.G; 1999. "The role of Green Water in food production in Sub-Saharan Africa". Background paper for the FAO internet-email conference on "Water for Food in sub-Saharan Africa", FAO, Rome.

Savenije, H.H.G. 2004. The importance of interception and why we should delete the term evapotranspiration from our vocabulary. *Hydrological Processes*, 18, 1507– 1511.

Savenije, H.H.G; van der Zaag, P. 2000. Conceptual framework for the management of shared river basins with special reference to the SADC and EU. *Water Policy*, 2, 9-45.

Sawunyama, T; Basima Busane, L; Chinoda, G; Twikirize, D; Love, D; Senzanje, A; Hoko, Z; Manzungu, E; Mangeya, P; Matura, N; Mhizha, A; Sithole, P. 2005. An integrated evaluation of a small reservoir and its contribution to improved rural livelihoods: Sibasa Dam, Limpopo Basin, Zimbabwe. In: Abstract volume, 6th WaterNet/WARFSA/GWP-SA Symposium, Swaziland, November 2005, p32. http://www.waternetonline.ihe.nl/challengeprogram/P02%20Sawunyama%20small%20dam.pdf

Sawunyama, T.; Senzanje, A.; Mhizha, A. 2006. Estimation of small reservoir storage capacities in Limpopo River Basin using geographical information systems (GIS) and remotely sensed surface areas: case of Mzingwane catchment. *Physics and Chemistry of the Earth*, 31, 935-943.

Schlenker, W; Lobell, D.B. 2010. Robust negative impacts of climate change on African agriculture. *Environmental Research Letters*, 5, 014010

Schulze R.E.; Hohls, B.C. 1993. A generic hydrological land cover and land use classification with decision support systems for use in models. *Proceedings, 6th South African National Hydrology Symposium*. University of Natal, Pietermaritzburg, Department of Agricultural Engineering, pp547-555.

Schulze R.E.; Lechler, N.L.; Hohls, B.C. 1995. Land cover and treatment. In: Schulze, R.E. (ed.) *The ACRU Theory Manual, Hydrology and Agrohydrology*. School of Bioresources Engineering and Environmental Hydrology, University of KwaZulu-Natal, South Africa. http://www.ukzn.ac.za/unp/beeh/acru/documentation/theory/AT_chapter6.PDF

Scibek, J.; Allen, D.M. 2006. Comparing modelled responses of two high-permeability, unconfined aquifers to predicted climate change. *Global Planetary Change*, 50, 50-62

Scoones, I; 1996. *Hazards and Opportunities; Farming Livelihoods in Dryland Africa: lessons from Zimbabwe*. Zed Books, London, 267p.

Scoones, I; Chaumba, J; Mavedzenge, B; Wolmer, W. 2012. The new politics of Zimbabwe's lowveld: Struggles over land at the margins. *African Affairs*, 111, 527-550.

Seibert, J. 1999. Regionalisation of parameters for a conceptual rainfall-runoff model. *Agric. For. Met.;* 98-99, 279-293.

Seibert, J.; 2000. Multi-criteria calibration of a conceptual runoff model using a genetic algorithm. *Hydrology and Earth System Sciences* 4, 215-224.

Seibert, J. 2002. *HBV light version 2 user's manual*. Environmental Assessment Department, SLU, Sweden, 32p.

Seibert, J.; Beven, K.J. 2009. Gauging the ungauged basin: how many discharge measurements are needed? *Hydrology and Earth Systems Sciences*, 13, 883-892.

Senzanje, A.; Samakande, I.; Chidenga, E.; Mugutso, D. 2003. Field irrigation practices and the performance of smallholder irrigation in Zimbabwe: case studies from Chakowa and Mpudzi irrigation schemes. *Journal of Agricultural Technology*, 5, 76-89.

Senzanje, A; Boelee, E; Rusere, S. 2011. Multiple use of water and water productivity of communal small dams in the Limpopo Basin, Zimbabwe. *Irrigation and Drainage Systems*, 22, 225–237. DOI: 10.1007/s10795-008-9053-7

Seyam, I.; Savenije, H.H.G.; Aerts, J.; Schepel, M. 2000. Algorithms for water resources distribution in international river basins. *Physics and Chemistry of the Earth, Part B*, 25, 309-314.

Shao, Q.; Li, Z.; Xu, Z. 2010. Trend detection in hydrological time series by segment regression with application to Shiyang River Basin. *Stochastic Environmental Research and Risk Assessment*, 24, 221-233.

Shah, T.; Burke, J.; Villholth, K. 2007. Groundwater: a global assessment of scale and significance. In: *Water for Food, Water for Life: A Comprehensive Assessment of Water Management in Agriculture*, ed., Molden, D., Earthscan, London and International Water Management Institute, Colombo, pp395-424.

Shields, F.D.Jr.; Simon, A.; Steffen, L.J. 2000. Reservoir effects on downstream river channel migration. *Environmental Conservation*, 27, 54-66.

Shoko, D.S.M.; Love, D. 2005. Gold panning law in Zimbabwe – challenges and contribution to integrated water resource management. In: *Water and Wastewater Management for Development Countries*, eds., Mathew, K.; Nhapi, I., IWA Water and Environmental Management Series, IWA Publishing, London, 499-512.

Shongwe, M.E.; van Oldenborgh, G.J.; van den Hurk, B.J.J.M.; de Boer, B.; Coelho, C.A.S.; van Aalst, M.K. 2009. Projected Changes in Mean and Extreme Precipitation in Africa under Global Warming. Part I: Southern Africa. *Journal of Climate* 22: 3819-3837.

Shumba, E. M.; Maposa, R. 1996. An evaluation of the performance of six smallholder irrigation schemes in Zimbabwe. *Irrigation and Drainage Systems*, 10, 355-366.

Sibanda, T.; Nonner, J.C.; Uhlenbrook, S. 2009. Comparison of groundwater recharge estimation methods for the semi-arid Nyamandhlovu area, Zimbabwe. *Hydrogeology Journal*, 17, 1427-1441.

Siemens, M.G.; Schaefer, D.M.; Vatthaeur, R.J. 1999. Rations for beef cattle. *University of Wisconsin-Madison Fact Sheet* A2387.

Sivakumar, M.Y.K.; Das, H.P.; Brunini, O. 2005. Impact of present and future climate variability and change on agriculture and forestry in the arid and semi-arid tropics. In: *Increasing Climate Variability and Change: Reducing the Vulnerability of Agriculture and Forestry*, eds., Salinger, J.; Sivakumar, M.Y.K.; Motha, R.P., Springer, Dordrecht. pp31-72.

Sivapalan M.; Takeuchi, K.; Franks, S.W.; Gupta, V.K.; Karambiri, H.; Lakshmi, V.; Liang X.; McDonnell, J.J.; Mendiondo, E.M.; O'Connell, P.E.; Oki, T.; Pomeroy, J.W.; Schertzer, D.; Uhlenbrook, S. and Zehe, E. 2003: IAHS Decade on Predictions in Ungauged Basins (PUB), 2003-2012: Shaping an Exciting Future for the Hydrological Sciences. *Hydrological Sciences Journal;* 48, 857-880.

Sithole, B.; 2001. Participation and stakeholder dynamics in the water reform process in Zimbabwe: the case of the Mazoe Pilot Catchment Board. *African Studies Quarterly*, 5.

Smith, R. E. 2004. Land tenure, fixed investment, and farm productivity: Evidence from Zambia's Southern Province. *World Development*, **32**, 1641-1661.

Smith, G.N.; Rethman, N.F.G. 2000. The influence of tree thinning on the soil water in a semi-arid savanna of southern Africa. *Journal of Arid Environments*, 44, 41-59.

SNV, 2001. *SNV Water Programme: RDC Focus*. Unpublished report, SNV Netherlands Development Organisation: Harare, 6p.

Stige, L.C.; Stave, J.; Chan, K.-S.; Ciannelli, L.; Pettorelli, N.; Glantz, M.; Herren, H.R.; Stenseth, N.C. 2006. The effect of climate variation on agro-pastoral production in Africa. *PNAS*, 103, 9, 3049–3053.

Stephens, M.A.; 1974. EDF statistics for goodness of fit and some comparisons. Journal of the American Statistical Association 69, 730-737.

Sunguro, S. 2001. The use of isotope techniques for assessment, development and management of groundwater resources in the Save River alluvial deposits. Unpublished report, Groundwater Branch, Zimbabwe National Water Authority, Harare.

Surveyor General of Zimbabwe; Forestry Commission, 1996. 1:250,000 Vegetation Map Series. Surveyor General of Zimbabwe, Harare.

Symphorian, G.R.; Madamombe, E.; van der Zaag, P. 2003. Dam operation for environmental water releases; the case of Osborne dam, Save catchment, Zimbabwe. *Physics and Chemistry of the Earth,* 28, 985-993.

Taigbenu, A.E; Ncube, M; Boroto R.J. 2005. Resources management in agriculture: convergence of needs and opportunities. In: 12th South African National Hydrology Symposium, Midrand, South Africa, September 2005, pp1-10.

Tate, E.L.; Freeman, S.N.; 2000. Three modelling approaches for seasonal streamflow droughts in southern Africa: the use of censored data. Hydrological Sciences Journal 45, 27-42.

Taylor, R.; Tindimugaya, C.; Barker, J.; Macdonald, D.; Kulabako, R. 2010. Convergent radial tracing of viral and solute transport in gneiss saprolite. *Ground Water*, 48, 284-294

Thirtle, C.; Atkins, J.; Bottomley, P.; Gonese, N.; Govereh, J.; Khatri, Y. 1993. Agricultural productivity in Zimbabwe, 1970-90. *The Economic Journal*, 103, 474-480.

Thompson, J.G.; Purves, W.D. 1978. A guide to the soils of Rhodesia. *Rhodesia Agricultural Journal Technical Handbook* No. 3. Chemistry and Soils Research Institute, Department of Research and Specialist Services. Government Printers, Salisbury, Rhodesia

Thompson, R.L. 1979. The geology of the area around Tuli, Mazunga and Gongwe. *Zimbabwe Geological Survey Short Report 40*

Thornton, P.K.; Kruska, R.L.; Henninger, N.; Kristjanson, P.M.; Reid, R.S.; Ateino, F.; Odero, A.N.; Ndegwa, T. 2002. *Mapping poverty and livestock in the developing world.* International Livestock Research Institute, Nairobi.

Thornton, P.K; Jones, P.G; Ericksen, P.J; Challinor, A.J. 2010. Agriculture and food systems in sub-Saharan Africa in a 4°C+ world. *Philosophical Transactions of the Royal Society A*, 369, 117-136.

Tilahun, K. 2007. The characterisation of rainfall in the arid and semi-arid regions of Ethiopia. *Water SA*, 32, 429-436.

Trenberth, K.E.; Jones, P.D.; Ambenje, P.; Bojariu, R.; Easterling, D.; Klein Tank, A.; Parker, D.; Rahimzadeh, F.; Renwick, J.A.; Rusticucci, M.; Soden, B.; Zhai, P. 2007. Observations: Surface and Atmospheric Climate Change. In: *Climate Change 2007: The Physical Science Basis. Contribution of Working Group I to the Fourth Assessment Report of the Intergovernmental Panel on Climate Change*, eds., Solomon, S.D.; Qin, M.; Manning, Z.; Chen, M.; Marquis, K.B.; Averyt, M.; Tignor, A.; Miller, H.L., Cambridge University Press, Cambridge. pp235-336.

Troch, P.A.; Dijksma, R.; van Lanen, H.A.J.; van Loon, E.; 2007. Towards improved observations and modelling of catchment-scale hydrological processes: bridging the gap between local knowledge and the global problem in ungauged catchments. In: *Predictions in Ungauged Basins: PUB Kick-off (Proceedings of the PUB Kick-off meeting held in Brasilia, 20–22 November 2002)*, IAHS Publication 309, pp173-185.

Tsiko, T.C.; Makurira, H.; Gerrits, A.M.J. and Savenije, H.H.G. 2008. Measuring forest floor and canopy interception in a savannah ecosystem (A case study of Harare, Zimbabwe). In: *Abstract volume, 9th WaterNet/WARFSA/GWP-SA Symposium, Johannesburg, South Africa*, November 2008, p9.

Tu, M. de Laat, P.J.M., Hall, M.J., de Wit, M.J.M. 2005. Precipitation variability in the Meuse basin in relation to atmospheric circulation. *Water Science and Technology*, 51, 5-14.

Tunhuma, N.; Kelderman, P.; Love, D.; Uhlenbrook, S. 2007. Environmental Impact Assessment of Small Scale Resource Exploitation: the case of gold panning in Zhulube Catchment, Limpopo Basin, Zimbabwe. In: *Abstract volume, 8th WaterNet/WARFSA/GWP-SA Symposium*, Livingstone, Zambia, November 2007, p47. **[Annex 1.14 of this thesis]**

Twidale, C.R. 1998. The missing link, planation surfaces and etch forms in southern Africa. In; *Geomorphological Studies in Southern Africa*, eds. Dardis, G.F.; Moon, B.P.; AA Balkema Publishers, Rotterdam, pp31-46

Twomlow, S.; Bruneau, P. 2000. Semi-arid soil water regimes in Zimbabwe. *Geoderma*, 95, 33-51.

Twomlow, S; Riches, C; O'Neill, D; Brookes, P; Ellis-Jones, J. 1999. Sustainable dryland smallholder farming in Sub-Saharan Africa. *Annals of Arid Zone*, 38, 93-135.

Twomlow, S.J.; Steyn, J.T.; Du Preez, C.C. 2006. Dryland farming in southern Africa. In: Dryland Agriculture, 2nd Edition. *Agronomy Monograph of the American Society of Agronomists*, 23, 769-836.

Twomlow, S.; Love, D.; Walker, S. 2008. The nexus between Integrated Natural Resources Management and Integrated Water Resources Management in Southern Africa. *Physics and Chemistry of the Earth*, 33, 889-898. **[Annex 1.7 of this thesis]**

Twomlow, S.; Urolov, J.C.; Jenrich, M.; Oldrieve, B. 2008. Lessons from the field – Zimbabwe's Conservation Agriculture Task Force. *Journal of SAT Agricultural Research*, 6, 1-11.

Tyson, P.D. 1986. *Climatic change and variability in southern Africa*. Oxford University Press, Oxford.

Uhlenbrook, S. 2007. Biofuel and water cycle dynamics: what are the related challenges for hydrological processes research? *Hydrological Processes* 21, 3647-3650.

Uhlenbrook, S.; Leibundgut, C.; 2002. Process-oriented catchment modelling and multiple-response validation. *Hydrological Processes* 16, 423-440.

Uhlenbrook, S.; Roser, S.; Tilch, N. 2004. Hydrological process representation at the meso-scale: the potential of a distributed, conceptual catchment model. *Journal of Hydrology*, 291, 278-296.

UN Millennium Project, 2005a. *Investing in Development: A Practical Plan to Achieve the Millennium Development Goals*. Earthscan, New York.

UN Millennium Project, 2005b. *Halving Hunger: It can be done. Report of the Task Force on Hunger*. Earthscan, New York.

Unganai, L.S.; Mason, S.J. 2002. Long-range predictability of Zimbabwe summer rainfall. *International Journal of Climatology*, 22, 1091-1103.

Vachaud, G.; Chen, T. 2002. Sensitivity of a large-scale hydrologic model to quality of input data obtained at different scales; distributed versus stochastic non-distributed modelling. *Journal of Hydrology*, 264, 101-112.

Van de Giesen, N.C.; Stomph, T.J.; de Ridder, N. 2000. Scale effects of Hortonian overland flow and rainfall-runoff dynamics in a West African catena landscape. *Hydrological Processes* 14, 165-175.

Van der Zaag, P. 2005. Integrated Water Resources Management: Relevant concept or irrelevant buzzword? A capacity building and research agenda for Southern Africa. *Physics and Chemistry of the Earth*, 30, 867-871.

Van der Zaag, P. 2009. Viewpoint–Water variability, soil nutrient heterogeneity and market volatility–Why sub-Saharan Africa's Green Revolution will be location-specific and knowledge-intensive. *Water Alternatives* 3, 154-160.

Van der Zaag, P.; Gupta, J. 2008. Scale issues in the governance of water storage projects. *Water Resources Research*, 44, W10417. DOI:10.1029/2007WR006364.

Van der Zaag, P; Savenije, H.H.G. 2000. Towards improved management of shared river basins: lessons from the Maseru Conference. *Water Policy*, 2, 47-63.

Van der Zaag, P.; Gupta, J.; Darvis, L.P. 2009. HESS Opinions" Urgent water challenges are not sufficiently researched". *Hydrology and Earth System Sciences*, 13, 905.

Van Koppen, B.; Mahmud, S. 1996. Women and Water-Pumps in Bangladesh: The Impact of Participation in Irrigation Groups on Women's Status. Pratical Action, Rugby, 256p.

Van Koppen, B.; Moriaty, P.; Boelee, E. 2006. Multiple-use water services to advance the millennium development goals. *IWMI Research Report*, 98, International Water Management Institute, Columbo.

Van Ty, T; Sunada, K; and Ichikawa, Y. 2011. A spatial impact assessment of human-induced intervention on hydrological regimes: a case study in the upper Srepok River basin, Central Highlands of Vietnam. *International Journal of River Basin Management*, 9, 103-116.

Vidal, A.; Van Koppen, B.; Love, D.; Ncube, B.; Blake, D. 2009. The Green to Blue Water Continuum: an approach to improve agricultural systems' resilience to water scarcity. *Stockholm Water Symposium, World Water Week*, Stockholm, Sweden, August 2009. http://www.worldwaterweek.org/documents/WWW_PDF/2009/tuesday/K16-17/Alain_Vidal_-_Green_to_Blue_Water_Continuum.pdf **[Annex 1.19 of this thesis]**

Vincent, V.; Thomas, R.G.; 1960. An agricultural survey of Southern Rhodesia: Part I: agro-ecological survey, Government Printer, Salisbury.

Von der Hayden, C.J.; New, M.G. 2003. The role of a dambo in the hydrology of a catchment and the river network downstream. *Hydrology and Earth Systems Sciences*, 7, 339-357.

Vörösmarty, C.J.; Green, P.; Salisbury, J.; Lammers, R.B. 2000. Global water resources: vulnerability from climate change and population growth. *Science*, 289, 284-288.

Vucetic, M. 1994. Cyclic threshold shear strains in soils. *Journal of Geotechnical Engineering*, 120, 2208-2229

Vukovic, M.; Soro, A. 1992 Determination of Hydraulic Conductivity of Porous Media from Grain-Size Composition. Water Resources Publications, Littleton, Colorado

Walker, J.; Dowling, T.; Veitch, S. 2006. An assessment of catchment condition in Australia. *Ecological Indicators*, 6, 205-214.

Ward, J.V.; Stanford, J.A. 1995. Ecological connectivity in alluvial river ecosystems and its disruption by flow regulation. *Regulated Rivers: Research and Management*, 11, 105-119.

Wiltshire, S.E.; 1986. Regional flood frequency analysis II: multivariate classification of drainage basins in Britain. *Hydrological Sciences Journal* 31, 335-346.

Woolridge, S.A.; Kalma, J.D.; Walker, J.P.; 2003. Importance of soil moisture measurements for inferring parameters in hydrologic models of low-yielding ephemeral catchments. *Environmental Modelling and Software* 18, 35-48.

Wolter, K. 2007. Multivariate ENSO Index. National Oceanic and Atmospheric Administration http://www.cdc.noaa.gov/people/klaus.wolter/MEI/table.html

Wolter, K., Timlin, M.S. 1998. Measuring the strength of ENSO events - how does 1997/98 rank? Weather, 53, 315-324.

Woltering, L; Ibrahim, A; Pasternak, D; Ndjeunga, J. 2011. The economics of low pressure drip irrigation and hand watering for vegetable production in the Sahel. *Agricultural Water Management*, 99, 67-73.

World Resource Institute. 2000. Cattle Density in Africa. http://www.earthtrends.wri.org/text/agriculture-food/map-246.html

Wright, E.P. 1992. The hydrogeology of crystalline basement aquifers in Africa. *Geological Society, London: Special Publications*, 66, 1-27.

Yu, B.; Neil, D. T. 1993. Long-term variations in regional rainfall in the south-west of Western Australia and the difference between average and high intensity rainfalls. *International Journal of Climatology*, 13, 77-88.

Zhang, X.; Zhang, L.; Zhao, J.; Rustomji, P.; Hairsine, P. 2008. Responses of streamflow to changes in climate and land use/cover in the Loess Plateau, China. *Water Resources Research*, 44, W00A07.

ZIMVAC (Zimbabwe Vulnerability Assessment Committee), 2012. *Rural Livelihoods Assessment: May 2012 Report*. Food and Nutrition Council, SIRDC, Harare.

Zinyama, L.; Whitlow, R. 2004. Changing patterns of population distribution in Zimbabwe. *GeoJournal*, 13, 365-384.

Zwane, N. Love, D.; Hoko, Z.; Shoko, D. 2006. Managing the impact of gold panners within the context of integrated water resources management planning: the case of the Lower Manyame Subcatchment, Zambezi Basin, Zimbabwe. *Physics and Chemistry of the Earth*, 31, doi:10.1016/j.pce.2006.08.024.

List of Symbols and Abbreviations

β	HBV partitioning function - see equation (4.9)
D	interception threshold (mm)
dV_d	daily volume error (-)
E	Soil evaporation and transpiration - see equation (4.8)
E_s	nett evaporation from saturated sand (m^3 d^{-1})
e_{10}	10 % elasticity index (-)
FC	Maximum soil moisture storage (mm)
G	Seepage from base of alluvial aquifer or deep percolation (m^3 d^{-1})
I	interception (mm d^{-1})
K	hydraulic conductivity (m d^{-1})
Kc	crop coefficient (-)
LP	soil moisture ratio threshold (-)
n	porosity (-)
P	rainfall (mm d^{-1})
P_{eff}	Effective rainfall (mm d^{-1})
$Perc$	Percolation (mm d^{-1})
Q_0	Overland flow (mm d^{-1})
Q_1	Discharge from saturated soil or shallow groundwater (mm d^{-1})
Q_2	Discharge from deep groundwater (mm d^{-1})
Q_c	Total discharge from catchment (mm d^{-1})
r	Correlation coefficient (-)
r^2	Determination coefficient (-)
S_y	Specific Yield (-)
$toRGR$	Moisture transferred to runoff generation routine (mm d^{-1})
U	coefficient of grain uniformity (-)
UZL	Threshold for start of overland flow (mm)
API	antecedent precipitation index (-)
CEO	Chief Executive Officer
CGIAR	Consultative Group on International Agricultural Research
CPWF	Challenge Program on Water and Food
DA	District Administrator
ENSO	El Niño – Southern Oscillation phenomenon
FAO	Food and Agricultural Organisaton of the United Nations
Fewsnet	Famine Early Warning System Network
GDP	Gross Domestic Product
GHB	General Head Boundary
HOWSIT	HBVx On WAFLEX Spreadsheet Information Tool
ICRISAT	International Crops Research Institute for the Semi-Arid Tropics
IPCC SRES	Intergovernmental Panel on Climate Change Special Report on Emission Scenarios
ITCZ	Inter Tropical Convergence Zone
IWRM	Integrated Water Resources Management
M27	Mnyabezi 27 catchment
MCC	Mzingwane Catchment Council
MDG	Millennium Development Goal

MEI	Multivariate El Niño – Southern Oscillation Index
MSH	Mushawe catchment
MUS	Multiple use system
NGO	Non-Governmental Organisation
PUB	Prediction in Ungauged Basins initiative
RDC	Rural District Council
SADC	Southern African Development Community
UBN	Upper Bengu catchment
VIDCO	Village Development Committee
WADCO	Ward Development Committee
ZIMVAC	Zimbabwe Vulnerability Assessment Committee
ZINWA	Zimbabwe National Water Authority

List of Figures

Annex 1: Additional publications arising from this PhD study [*]

Annex 1.1.   **Alluvial aquifers in the Mzingwane Catchment: their distribution, properties, current usage and potential expansion**

Moyce, W., Mangeya, P., Owen, R. and Love, D. 2006. Alluvial aquifers in the Mzingwane Catchment: their distribution, properties, current usage and potential expansion. *Physics and Chemistry of the Earth*, **31**, 988-994. doi:10.1016/j.pce.2006.08.013.

This paper received the Phaup award from the Geological Society of Zimbabwe in 2006.

The Mzingwane River is a sand filled channel, with extensive alluvial aquifers distributed along its banks and bed in the lower catchment. LandSat TM imagery was used to identify alluvial deposits for potential groundwater resources for irrigation development. On the false colour composite band 3, band 4 and band 5 (FCC 345) the alluvial deposits stand out as white and dense actively growing vegetation stands out as green making it possible to mark out the lateral extent of the saturated alluvial plain deposits using the riverine fringe vegetation signature. Alluvial aquifers form ribbon shaped aquifers extending along the channel and reaching over 20 km in length in some localities and are enhanced at lithological boundaries. These alluvial aquifers extend laterally outside the active channel. The alluvial aquifers are more pronounced in the Lower Mzingwane where the river gradient is gentler and allows for more sediment accumulation. Estimated water resources potential or those sites which were identified in this study comes to 15,665,000 m^3 for aquifers in the river channels and for aquifers on the alluvial plains to 22,230,000 m^3. Such a water resource potential can support irrigation of 1,567 ha for the river channel aquifers and 2,223 ha for the plain aquifers.

Currently some of these aquifers are being used to provide water for domestic use, livestock watering and dip tanks, commercial irrigation and market gardening. The water quality of the aquifers in general is fairly good due to regular recharge and flushing out of the aquifers by annual river flows and floodwater. Water salinity was found to increase significantly by the end of the dry season, and this effect was more pronounced in water abstracted from wells on the alluvial plains.

Annex 1.2.   **Effects of selected dams on river flows of Insiza River, Zimbabwe**

Kileshye Onema, J.-M., Mazvimavi, D., Love, D. and Mul, M. 2006. Effects of dams on river flows of Insiza River, Zimbabwe. *Physics and Chemistry of the Earth*, **31**, 870-875. doi:10.1016/j.pce.2006.08.022

This paper examines effects of three dams on flow characteristics of Insiza River on which they are located. The storage capacities of these dams varies from an equivalent of 48–456% of the mean annual runoff. Mean annual runoff and annual maximum flood flows have not been modified by the presence of these dams. The

---

[*] Papers with students and collaborative syntheses

average number of days per year without runoff had decreased downstream of two dams. A comparison was made of flow duration curves at sites upstream and downstream of the selected dams. Significant differences were detected between the flow duration curves of upstream and downstream sites. Exceedance frequencies of low flows had decreased downstream of two dams, while these had increased downstream of the other dam. The study recommends development of operating rules for these dams that will ensure that changes detected in low flows do not adversely affect instream flow requirements.

Annex 1.3.   Soil-water conservation and other rainwater harvesting strategies in the semi-arid Mzingwane Catchments, Limpopo Basin, Zimbabwe

Mupangwa, W., Love, D. and Twomlow, S. 2006. Soil-water conservation and other rainwater harvesting strategies in the semi-arid Mzingwane Catchments, Limpopo Basin, Zimbabwe. *Physics and Chemistry of the Earth*, **31**, 893-900. doi:10.1016/j.pce.2006.08.042

Various soil water management practices have been developed and promoted for the semi-arid areas of Zimbabwe. These include a variety of infield crop management practices that range from primary and seconday tillage approaches for crop establishment and weed management through to land forming practices such as tied ridges and land fallowing. Tillage methods evaluated in this study include deep winter ploughing, no till tied ridges, modified tied ridges, clean and mulch ripping, and planting basins. Data collected from the various trials since the 1990's show that mulch ripping and other minimum tillage practices consistently increased soil water content and crop yields compared to traditional spring ploughing. Trial results also showed higher soil loss from conventionally ploughed plots compared to plots under different minimum tillage practices.

Annex 1.4.   Impact and sustainability of drip irrigation kits in the semi-arid Limpopo Basin, Zimbabwe

Moyo, R., Love, D., Mul, M., Twomlow, S. and Mupangwa, W. 2006. Impact and sustainability of low-head drip irrigation kits, in the semi-arid Gwanda and Beitbridge Districts, Mzingwane Catchment, Limpopo Basin, Zimbabwe. *Physics and Chemistry of the Earth*, **31**, 885-892. doi:10.1016/j.pce.2006.08.020

Resource-poor smallholder farmers in the semi-arid Gwanda and Beitbridge districts face food insecurity on an annual basis due to a combination of poor and erratic rainfall (average 500 mm a^{-1} and 345 mm a^{-1} respectively, for the period 1970 to 2003) and technologies inappropriate to their resource status. This impacts on both household livelihoods and food security. In an attempt to improve food security in the catchment a number of drip kit distribution programmes have been initiated since 2003 as part of an on-going global initiative aimed at 2 million poor households per year. A number of recent studies have assessed the technical performance of the drip kits in-lab and in-field. In early 2005 a study was undertaken to assess the impacts and sustainability of the drip kit programme. Representatives of the NGOs, local government, traditional leadership and agricultural extension officers were interviewed. Focus group discussions with beneficiaries and other villagers were held at village level. A survey of 114 households was then conducted in two districts, using a questionnaire developed from the output of the interviews and focus group discussions.

The results from the study showed that the NGOs did not specifically target the distribution of the drip kits to poor members of the community (defined for the purpose of the study as those not owning cattle). Poor households made up 54 % of the beneficiaries. This poor targeting of vulnerable households could have been a result of conditions set by some implementing NGOs that beneficiaries must have an assured water source. On the other hand, only 2 % of the beneficiaries had used the kit to produce the expected 5 harvests over the 2 years, owing to problems related to water shortage, access to water and also pests and diseases. About 51 % of the respondents had produced at least 3 harvests and 86 % produced at least 2 harvests. Due to water shortages during the dry season 61% of production with the drip kit occurred during the wet season. This suggests that most households use the drip kits as supplementary irrigation. Conflicts between beneficiaries and water point committees or other water users developed in some areas especially during the dry season. The main finding from this study was that low cost drip kit programs can only be a sustainable intervention if implemented as an integral part of a long-term development program, not short-term relief programs and the programme should involve a broad range of stakeholders. A first step in any such program, especially in water scarce areas such as Gwanda and Beitbridge, is a detailed analysis of the existing water resources to assess availability and potential conflicts, prior to distribution of drip kits.

Annex 1.5.   Involvement of stakeholders in the water quality monitoring and surveillance system: the case of Mzingwane Catchment

Nare, L., Love, D. and Hoko, Z. 2006. Involvement of stakeholders in the water quality monitoring and surveillance system: the case of Mzingwane Catchment. *Physics and Chemistry of the Earth*, **31**, 707-712. doi:10.1016/j.pce.2006.08.037

Stakeholder participation is viewed as critical in the current water sector reforms taking place in the region including Zimbabwe. Zimbabwean policies and legislation encourage stakeholder participation. A study was undertaken to determine the extent of stakeholder participation in water quality monitoring and surveillance at the operational level, and also to assess indigenous knowledge and practices in water quality monitoring. 241 questionnaires were administered in Mzingwane Catchment, the portion of the Limpopo Basin that falls within Zimbabwe. The focus was on small users in rural communities, whose experiences were captured using a questionnaire and some focus groups discussions. Extension workers, farmers and NGOs and relevant sector government ministries and departments were also interviewed and a number of workshops held.

Results indicate that there is very limited stakeholder participation although there are adequate structures and organisations to support this. For the Zimbabwe National Water Authority, stakeholders are the paying permit holders, who they give feedback after analysis of samples. The Ministry of Health and Child Welfare generally only releases information to rural communities when it is deemed necessary for their welfare. There are no guidelines on how a dissatisfied member of the public can raise a complaint - although some stakeholders carry such complaints to Catchment Council meetings. It was found out that there are many useful indigenous knowledge and practices used by the communities of the area, such knowledge is based on smell, taste, colour and odour perceptions. Residents are generally more concerned about the physical parameters than the bacteriological quality of water. They are aware of what causes water pollution and the effects of pollution on human health, crops, animals and aquatic ecology. They have ways of preventing pollution and interventions to take when a source of water is polluted, such as boiling water for human consumption, laundry and bathing, or abandoning a water source in extreme cases. Stakeholder participation and ownership of resources needs to be encouraged through participatory planning, and integration between the three government departments (water, environment and health). Local knowledge systems could be integrated into the formal water quality monitoring systems, in order to complement the conventional monitoring networks.

Annex 1.6.   **Potential water supply of a small reservoir and alluvial aquifer system in southern Zimbabwe**

De Hamer, W. Love, D., Owen, R.J.S., Booij, M.J. and Hoekstra, A. 2008. Potential water supply of a small reservoir and alluvial aquifer system in southern Zimbabwe. *Physics and Chemistry of the Earth*, **33**, 633-639. doi:10.1016/j.pce.2008.06.056

Groundwater use by accessing alluvial aquifers of non-perennial rivers can be an important additional water resource in the semi-arid region of southern Zimbabwe. The research objective of the study was to calculate the potential water supply for the upper-Mnyabezi catchment under current conditions and after implementation of two storage capacity measures. These measures are heightening the spillway of the 'Mnyabezi 27' dam and constructing a sand storage dam in the alluvial aquifer of the Mnyabezi River. The upper-Mnyabezi catchment covers approximately 22 km^2 and is a tributary of the Thuli River in southern Zimbabwe. Three coupled models are used to simulate the hydrological processes in the Mnyabezi catchment. The first is a rainfall-runoff model, based on the SCS-method. The second is a spreadsheet-based model of the water balance of the reservoir. The third is the finite difference groundwater model MODFLOW used to simulate the water balance of the alluvial aquifer.

The potential water supply in the Mnyabezi catchment under current conditions ranges from 2,107 m^3 (5.7 months) in a dry year to 3,162 m^3 (8.7 months) in a wet year. The maximum period of water supply after implementation of the storage capacity measures in a dry year is 2,776 m^3 (8.4 months) and in a wet year the amount is 3,617 m^3 (10.8 months). The sand storage dam can only be used as an additional water resource, because the storage capacity of the alluvial aquifer is small. However, when an ephemeral river is underlain by a larger alluvial aquifer, a sand storage dam is a promising way of water supply for smallholder farmers in southern Zimbabwe.

Annex 1.7. **The Nexus between Integrated Natural Resources Management and Integrated Water Resources Management in Southern Africa**

Twomlow, S., Love, D. and Walker, S. 2008. The nexus between Integrated Natural Resources Management and Integrated Water Resources Management in Southern Africa. *Physics and Chemistry of the Earth*, **33**, 889-898. doi:10.1016/j.pce.2008.06.044

The low productivity of smallholder farming systems and enterprises in the drier areas of the developing world can be attributed mainly to the limited resources of farming households and the application of inappropriate skills and practices that can lead to the degradation of the natural resource base. This lack of development, particularly in southern Africa, is of growing concern from both an agricultural and environmental perspective. To address this lack of progress, two development paradigms that improve land and water productivity have evolved, somewhat independently, from different scientific constituencies. One championed by the International Agricultural Research constituency is Integrated Natural Resource Management (INRM), whilst the second championed predominantly by Environmental and Civil Engineering constituencies is Integrated Water Resources Management (IWRM). As a result of similar objectives of working towards the millennium development goals of improved food security and environmental sustainability, there exists a nexus between the constituencies of the two paradigms, particularly in terms of appreciating the lessons learned. In this paper lessons are drawn from past INRM research that may have particular relevance to IWRM scientists as they re-direct their focus from blue water issues to green water issues, and vice-versa. Case studies are drawn from the management of water quality for irrigation, green water productivity and a convergence of INRM and IWRM in the management of gold panning in southern Zimbabwe. One point that is abundantly clear from both constituencies is that 'one-size-fits-all' or silver bullet solutions that are generally applicable for the enhancement of blue water management/formal irrigation simply do not exist for the smallholder rainfed systems.

Annex 1.8. **Modelling upstream-downstream interactions using a spreadsheet-based water balance model: two case studies from the Limpopo basin**

Love, D., Khosa, S., Mul, M., Uhlenbrook, S. and van der Zaag, P. 2008. Modelling upstream-downstream interactions using a spreadsheet-based water balance model: two case studies from the Limpopo basin. In: Humphreys, E., Bayot, R.S., van Brakel, M., Gichuki, F., Svendsen, M., Wester, P., Huber-Lee, A., Cook, S., Douthwaite, B., Hoanh, C.T., Johnson, N., Nguyen-Khoa, S., Vidal, A. and MacIntyre, I. (eds.). *Fighting Poverty Through Sustainable Water Use: Proceedings of the CGIAR Challenge Program on Water and Food 2nd International Forum on Water and Food, Addis Ababa, Ethiopia, November 10 – 14 2008*, **IV**. The CGIAR Challenge Program on Water and Food, Colombo, pp15-21. [ISBN 9789299005323]

Sharing scarce surface water resources requires tools for water allocation. Water balance modelling can be used to analyse upstream-downstream interactions, dam management options and water allocation and development options. The spreadsheet-based model WAFLEX provides such a modelling tool. In this study, the lower Mzingwane and Thuli river basins, Limpopo Basin, Zimbabwe, are modelled in WAFLEX in order to assess existing upstream-downstream interactions and future development and allocation options. The lower Mzingwane river is managed through releases from Zhovhe Dam. The downstream river reaches, via an alluvial aquifer, supply water to an important agro-industry (growing and processing citrus and other crops) and to the border town of Beitbridge. The WAFLEX model shows how, through management of the dam and planning of releases, these water resources can be better shared to also provide irrigation water for smallholder farmers in the communal lands along the river, where soils are poor and rainfall low and erratic. The Thuli river basin is dominated by the Thuli-Makwe Dam, which supplies water for smallholder irrigation, and the Mtshabezi dams, that provide urban water supply. An inter-basin water transfer is proposed from Mtshabezi Dam to Bulawayo, the second largest city in Zimbabwe. The WAFLEX model shows how the water can be equitably shared with minimal negative impacts.

Annex 1.9.   Methodologies and case studies for investigating upstream-downstream interactions of rainwater harvesting in the Limpopo Basin

Ncube, B., Magombeyi, M., Munguambe, P., Mupangwa, W. and Love, D. 2009. Methodologies and case studies for investigating upstream-downstream interactions of rainwater water harvesting in the Limpopo Basin. *In*: Humphreys, L (ed). *Proceedings of the Workshop on Increasing the Productivity and Sustainability of Rainfed Cropping Systems of Poor, Smallholder Farmers, Tamale, Ghana, 22-25 September 2008*, The CGIAR Challenge Program on Water and Food, Colombo, 209-221. [ISBN 9789299005347]

Rainwater harvesting (RWH) is a promising technology for increasing water availability for crop production of smallholder farmers in the semi-arid regions of the Limpopo Basin. A few studies on rainwater harvesting have been conducted in the basin at small plot and farmer field scales. Results from Mozambique, Zimbabwe and South Africa indicate substantial benefits to crops grown using a range of rainwater harvesting techniques. However, there have been no catchment and basin level studies to investigate the impacts of wide scale adoption at these levels. A methodology flow chart is proposed for systematically investigating the impacts of out-scaling of these in-field and ex-field rainwater harvesting techniques. The method proposes an analysis of levels of adoption to help identify optimum levels that will maximize land and water productivity while minimizing negative hydrological and ecological impacts at catchment or basin scales.

Annex 1.10.   Integrated water resource management (IWRM) for improved rural livelihoods: managing risk, mitigating drought and improving water productivity in the water-scarce Limpopo Basin

Ncube, B., Manzungu, E., Love, D., Magombeyi, M., Gumbo, B. and Lupankwa, K. 2010. The Challenge of Integrated Water Resource Management for Improved Rural Livelihoods: Managing Risk, Mitigating Drought and Improving Water Productivity in the Water Scarce Limpopo Basin. *Challenge Program on Water and Food Project Report* **17**.

It is increasingly understood that integrated water resource management (IWRM) is required, not only to balance water for food and nature, but also to unlock paths to sustainable development. A global hotspot area in terms of water for food and improved livelihoods is in the poverty stricken rural areas of water scarce semi-arid tropics, such as in the Limpopo basin. Here, translating IWRM from concept to action still remains largely undone. Water policy and institutions are embedded in a conventional blue water framework, mainly concerned with (runoff-based) water supply for irrigation, domestic and industrial use. This water resource strategy has limitations. Blue water resources for irrigation are over-committed in the Limpopo basin, while the bulk of agricultural produce sustaining lives of resource poor farmers originates from green water flows in rainfed crop and livestock production.

Integrated Water Resources Management (IWRM) is a systems approach to water management, based on the principle of managing the full water cycle (Twomlow *et al.*, 2008b). Green water is the source of runoff and percolation – and thus of blue water. The fundamental principles of IWRM are: (i) water is a vulnerable and finite resource requiring sustainable management, (ii) water is a special economic good, (iii) water management requires a participatory approach and (iv) sustainable water management requires the promotion of gender equity (ICWE, 1992; Savenije, 2002). The improvement in resilience that the IWRM approach can impart to rural livelihood systems has been shown by a series of case studies in the Limpopo Basin. Community or catchment water resource assessments must become an essential precursor to food security interventions, due to the convergence of water scarcity and food scarcity, and the constraints that water resource availability impose on development initiatives in basins such as the Limpopo (Love *et al.*, 2006b, 2010).

Access to green water in rainfed farming can be improved through a package of conservation agriculture techniques. Conservation tillage methods, such as planting basins, help to concentrate rainfall that falls in the field into the root zone of the crops and decreases runoff out of the field (Ncube et al., 2009). Best results are obtained when such methods are combined with fertility improvements such as manure, or micro-dosing with nitrogen fertilizer or with measures such as mulching that improve the use of water by crops and also decrease evaporation (Mupangwa, 2009). Yield improvements in rainfed farming translate very quickly into major improvements in green water productivity (Ncube *et al.*, 2007; Rockström *et al.*, 2007). The farming system's resilience is thus raised without industrial scale interventions.

Supplementary irrigation, using micro-catchment or runoff farming incorporates small-scale utilization of blue water into rainfed farming. It thus represents a nexus between rainfed and irrigated farming and conjunctive use of green and blue water. Studies in the Limpopo Basin (Mwenge Kahinda *et al.*, 2007; Magombeyi *et al.*, 2008) have shown that there is a substantial yield gap which supplementary irrigation can bridge. This is particularly the case especially during years with dryspells during the growing season, when conventional rainfed agriculture may fail completely.

A multi-stakeholder approach to decision-making, especially where gendered, builds resilience as negotiation processes between users result in new institutions, or new roles for existing institutions, such as school boards which take over borehole management. Such institutions often evolve and revolve around specific infrastructure (Mabiza *et al.* 2006). At the same time, these community-based institutions need linkage to formal water management structures (Dzingirai and Manzungu, 2009).

It should be emphasized that, as a network, WaterNet itself is a partnership organization. The Trust oversees, the secretariat coordinates but most activities are delivered by WaterNet members and partners. A key aspect of this project for WaterNet was building solid partnerships outside the university sector: with CG centres, government departments and NGOs. Many of these partnerships have outlasted the project. Individual project members built up their own partnerships, especially universities with CG centres, government departments and NGOs. This has led to cross-fertilisation and benefits to university curriculum, other research initiatives and so on. The methodology used by WaterNet as a network in developing the concept note and proposal, assembling the PN17 partnership and managing the project is serving as an excellent example as a way for us to facilitate our members to access international research programs. WaterNet is establishing other research projects in the same fashion.

Annex 1.11.   Case studies of groundwater – surface water interactions and scale relationships in small alluvial aquifers

Love, D. de Hamer, W., Owen, R.J.S., Booij, M.J., Uhlenbrook, S., Hoekstra, A. and van der Zaag, P. 2007. Case studies of groundwater – surface water interactions and scale relationships in small alluvial aquifers. In: *Abstract volume, 8th WaterNet/WARFSA/GWP-SA Symposium*, Lusaka, Zambia, November 2007, p21.

An alluvial aquifer can be described as a groundwater system, generally unconfined, that is hosted in laterally discontinuous layers of gravel, sand, silt and clay, deposited by a river in a river channel, banks or flood plain. In semi-arid regions, streams that are associated with alluvial aquifers tend to vary from discharge water bodies in the dry season, to recharge water bodies during certain times of the rainy season or when there is flow in the river from managed reservoir releases. Although there is a considerable body of research on the interaction between surface water bodies and shallow aquifers, most of this focuses on systems with low temporal variability. In contrast, highly variable, intermittent rainfall patterns in semi-arid regions have the potential to impose high temporal variability on alluvial aquifers, especially for small ones. Small alluvial aquifers are here understood to refer to aquifers on rivers draining a meso-catchment (scale of approximately $10^1 - 10^3$ km^2). Whilst these aquifers have lower potential storage than larger ones, they may be easier to access for poor rural communities – the smaller head difference between the riverbed and the bank can allow for cheap manual pumps. Thus, accessing small alluvial aquifers for irrigation represents a possibility for development for smallholder farmers. The aquifers can also provide water for livestock and domestic purposes. However, the speed of groundwater depletion after a rain event is often poorly understood. In this study, three small alluvial aquifers in the Limpopo Basin, Zimbabwe, were studied: (i) upper Bengu catchment, 8 km^2 catchment area on a tributary of the Thuli River, (ii) Mnyabeze 27 catchment, 22 km^2 catchment area on a tributary of the Thuli River, and (iii) upper Mushawe catchment, 350 km^2 catchment area on a tributary of the Mwenezi River. All three are ephemeral rivers. In each case, the hydrogeological properties of the aquifer were studied; the change in head in the aquifer was monitored over time, as well as any surface inflows. Results from each case are compared showing that scale imposes a lower limit on alluvial aquifer viability, with the shallowness of the Bengu aquifer (0.3 m) meaning it has effectively no storage potential. The much higher storage of the

Mushawe aquifer, as well as the longer period of storage after a flow event, can be assigned partially to scale and partially to the geological setting.

Annex 1.12.   Evaluating the effect of water demand scenarios on downstream water availability in Thuli river basin, Zimbabwe

Khosa, S., Love, D. and Mul., M. 2008. Evaluation of the effects of different water demand scenarios on downstream water availability: The case of Thuli river basin. In: *Abstract Volume, 9th WaterNet/WARFSA/GWP-SA Symposium*, Johannesburg, South Africa, October 2008, p78. [online]

Thuli river basin, in south western Zimbabwe, is situated in a semi-arid area, where surface water resource availability is a constraint. There is intensive use of blue water in the upper catchment more than its lower reaches. The paper presents the evaluation of the effects of upstream water demand scenarios on downstream users in the river basin. A model was applied as a tool to simulate the effects.

The impacts of different water demand scenarios on downstream water availability were evaluated. The water demand scenarios used were based on government recommendations and future plans on water resources development, drought risk mitigation, implementation of environmental water requirement and implementing inter basin transfer (IBT) to Bulawayo, the second largest city in Zimbabwe. The study showed that implementing IBT will increase water shortages for downstream users while enforcement of environmental water requirements, implementation of government plans on water resources development in the catchment and drought risk reduction; decreases water shortages for downstream users. It is therefore clear that while the IBT is an important development for Bulawayo, the river basin management of the Thuli river basin requires a holistic approach. Downstream users in the form of domestic and agricultural users should be considered while allocating water for the IBT.

Annex 1.13.   Evaluation of the Groundwater potential of the Malala alluvial aquifer, Lower Mzingwane river, Zimbabwe

Masvopo, T., Love, D. and Makurira, H. 2008. Evaluation of the groundwater potential of the Malala Alluvial Aquifer, Lower Mzingwane River, Zimbabwe. In: Abstract Volume, 9th WaterNet/WARFSA/GWP-SA Symposium, Johannesburg, South Africa, October 2008, p7

The largest river in the semi-arid southwest of Zimbabwe, the Mzingwane River, is ephemeral and thus can only supply water for a limited period of time during the year. This limited temporal availability of surface water can be mitigated through accessing water stored in the river bed: the alluvial aquifer. This study evaluated groundwater resources at a local scale by characterizing the Malala alluvial aquifer, which covers a stretch of 1000 m of the Mzingwane river and is on average 200 m wide. The aquifer is recharged naturally by flood events during the rainy season and artificially by managed dam releases from Zhovhe dam during the dry season. The Malala site was selected from geological mapping and resistivity studies. The site shows indications of deeper sand layers and hence would be expected to have a higher potential of storing more groundwater. Piezometers were installed in the river channel to monitor the water level fluctuations in the alluvial aquifer. Water samples were collected from Zhovhe dam, Mazunga area and Malala alluvial aquifer in order to analyse the major ion chemistry of the water at the aquifer and at the source of recharge. A piper diagram analysis showed that the water in the alluvial aquifer can be classified as sulphate water with no dominant metals. The water is also of a low sodium hazard and can therefore be used for irrigation without posing much risk to the compaction of soils. Laboratory and field tests gave an average porosity of 39 %, hydraulic conductivity of 37 m day^{-1}, specific yield value of 5.4 % and the slope of the aquifer was measured as 0.38 %. Resistivity surveys showed that the alluvial aquifer has an average depth of 13.4 m. The bedrock is metamorphic rock mainly tonalitic and granodioritic gneisses which have been intruded by a dolerite dyke. Water level observations from the installed piezometers indicated that the water levels dropped on average by 0.75 m within 100 days after the observed dam release. After any flow event in the Mzingwane River, approximately 135 ML of water per km stretch of the river is available for use by the communal farmers at Malala, with the potential of irrigating at least 13.5 ha a^{-1}. The frequency of such flows suggests the abstraction would be possible throughout the year. The alluvial aquifer can thus store a significant amount of water and has the potential to sustain both irrigation water supply throughout the year.

Annex 1.14.   **Environmental Impact Assessment of Small Scale Resource Exploitation: gold panning in Zhulube Catchment, Limpopo Basin**

Tunhuma, N., Kelderman, P., Love, D. and Uhlenbrook, S. 2007. Environmental Impact Assessment of Small Scale Resource Exploitation: the case of gold panning in Zhulube Catchment, Limpopo Basin, Zimbabwe. In: *Abstract volume, 8th WaterNet/WARFSA/GWP-SA Symposium*, Livingstone, Zambia, November 2007, p47. [online]

In sub-Saharan Africa most of the population is poverty-stricken and living in the rural areas. These people support their livelihoods by exploiting the natural resources in their vicinity. This study assesses the impacts of one such small-scale natural resource exploitation: gold panning in the Zhulube catchment in the Limpopo basin, Zimbabwe. Environmental impact assessment was done using the pressure-state-impact-response approach. The state was evaluated based on the researcher observation; water quantity was estimated using rainfall and siltation was estimated using two weirs in the catchment together with suspended solids in river water. Water quality was based on chemical water analyses of samples collected from the rivers in the catchment. A survey and informal interviews where carried out to assess the response. The results show small scale resources exploitation in Zhulube catchment have negative impacts on the environment in general and water resources in particular. These activities cause land clearance, erosion, sedimentation and introduction of pollutants among other environmental impacts. The most significant driver of environmental degradation, gold panning, was observed to cause an increase in sediment, an elevation of sulphates entering water bodies and an introduction of the toxic metal mercury into the aquatic environment. Apart from limited enforcement of and compliance with national law, poor resource use practises, a lack of sense of ownership as well as the need to generate livelihoods among users are responsible for the generation of these impacts. It is therefore recommended that illegal forms of small scale resource exploitation such as gold panning should be formalised. Local communities should also be involved in policy making and environmental protection. Furthermore, a continuous and systematic environmental monitoring system should be set up. This system would then act as the basis of decision making in areas of small scale resource exploitation.

Annex 1.15.   Livelihood challenges posed by water quality in the Mzingwane and Thuli river catchments, Zimbabwe

Love, D., Moyce, W. and Ravengai, S. 2006. Livelihood challenges posed by water quality in the Mzingwane and Thuli river catchments, Zimbabwe. *7th WaterNet/WARFSA/GWP-SA Symposium*, Lilongwe, Malawi, November 2006. [online]

Most strategies to improve rural livelihoods, such as upgrading agriculture, improving water supply and sanitation or industrialisation require increased water supply. In semi-arid areas, water resource availability is an obvious constraint. Water chemistry, which is a major control on the quality of water resources, can also be a constraint. Human health requires water that is both safe to drink (especially low in metals and nitrates) and palatable. Water for livestock must meet similar requirements. Water for irrigation – a major livelihood intervention in low rainfall areas – must have low salinity levels in order to avoid clogging delivery systems and must also be safe for plants. On this basis, many livelihood interventions require a water quality assessment. This study looked at the possible impact of ambient water quality on livelihood interventions in two river catchments in low rainfall areas of south-western Zimbabwe. 36 water samples were collected from rivers and alluvial aquifers in the Mzingwane and Thuli river catchments. The samples were analysed for alkalinity, cadmium, calcium, chloride, copper, fluoride, iron, magnesium, manganese, nickel, nitrate, nitrite, pH, phosphate, potassium, sodium, sulphate, total dissolved solids and zinc.

Results show that the ambient river water quality in the upstream tributaries is generally satisfactory, although some parameters such as arsenic and antimony were not analysed for. Water in some river reaches show high (although not toxic) metal levels. This is partly an ambient condition and partly due to pollution from mining. The latter can be better managed to reduce any risk to human and livestock health. The principle challenge encountered is that many alluvial aquifers in the downstream catchments, especially smaller aquifers and those on river bank flood plains, are characterised by high levels of sodium and chloride. This is an ambient condition, related to the geology of the aquifers, and threatens irrigated agriculture with equipment or crop failure. It necessitates the characterisation of boreholes and other water points as suitable or unsuitable for irrigation, prior to interventions such as drip kit distribution. Some alluvial aquifers showed elevated nitrate levels, which is a serious risk if the water is used for human consumption. Awareness campaigns are required to inform residents of those river reaches and aquifers where water is unsuitable for human consumption.

Annex 1.16.   A model for reservoir yield under climate change scenarios for the water-stressed City of Bulawayo, Zimbabwe

Moyo, B., Madamombe, E. and Love, D. 2005. A model for reservoir yield under climate change scenarios for the water-stressed City of Bulawayo, Zimbabwe. In: Abstract Volume, 6th WaterNet/WARFSA/GWP-SA Symposium, Swaziland, November 2005, p38. [online]

The City of Bulawayo, the second largest city in Zimbabwe (population 981,000), although located in the Zambezi Basin, sources its water mainly from five reservoirs located in the Upper Mzingwane Subcatchment, Limpopo Basin. The reservoirs have failed to store their expected 4 % yields. The city has frequently implemented water rationing or lesser restrictions. The city is currently under stringent water rationing since the total runoff into the water supply reservoir was low. A study was undertaken to determine if reservoir yields were falling, and possible causes thereof.

Examination of 30-year rainfall and runoff records showed declining precipitation and runoff on the five year moving averages. Dam yields, determined by Yield 200 model, have been declining over the same period, from 131.3 Mm^3 in 1980 to 67.90 Mm^3 in 2005. The phenomenon, which might influence such a trend, is global climate change. Future yields of the reservoirs were estimated by two methods. The yield was projected from the 30-year trend to year 2030 using the best-fit line. This formed one scenario, while two additional projections were determined using two scenarios of runoff and precipitation decline, predicted from IPCC SRES emissions scenarios. Both of these predicted less reduced yields by 2030 (45.04 Mm^3 and 41.72 Mm^3) than the yield projection from current data (67.90 Mm^3), suggesting the possibility that the impact of climate change in southern Zimbabwe may be higher than predicted by global models.

Unrestricted water demand for the city is projected to 83.70 Mm^3 of raw water by year 2030. This is far above even the most optimistic yield projection. It can therefore be concluded that the City should have determined additional sources of water already. The state-owned Mtshabezi dam and the Nyamandlovu aquifer connection could provide an increased total water supply in the short-term, but longer-term solutions, water demand management or additional sources, or both are required.

Annex 1.17.   Effects of grazing management on rangeland soil hydrology, Insiza, Zimbabwe

Ngwenya, P.T., Love, D., Mhizha, A. and Twomlow, S. 2006. Effects of grazing management on rangeland soil hydrology, Insiza, Zimbabwe. 7th WaterNet/WARFSA/GWP-SA Symposium, Lilongwe, Malawi, November 2006. [online]

Overgrazing by livestock has caused major changes in the productivity and composition of rangeland vegetation in Africa and other continents. The main problem stems from the fact that the carrying capacity of rangelands is low as a result of low vegetation cover, and is decreasing with range degradation. In Insiza, the present livestock density of 3.4 Ha/LU exceeds the recommended carrying capacity of 8 Ha/LU; which is the reason contributing to overgrazing of rangelands by livestock. This has an impact on land degradation which affects the rangeland hydrology. Thus the aim of this study was to determine the physio-hydrological responses of soil to different intensities of livestock grazing and land management by comparing the effect of uncontrolled grazed land, fenced off (ungrazed) land and stone lined grazed treatments. There is a need to understand the hydrology of rangeland so as to propose ways of improving carrying capacity of rangeland. The study was carried out over the 2005-2006 rainy season. Two range sites were chosen: one with clay soil and one with clay loam soil. Each site had three different treatments: (i) fenced off to prevent grazing, (ii) stone lined grazed and (iii) uncontrolled grazed treatments. Infiltration was measured with the use of tension infiltrometer and soil moisture was measured with the use of Time Domain Reflectrometer (TDR) soil moisture meter. Plant biomass was measured at the end of the season.

The results show that there is a significant different in infiltration rate and soil moisture among the two sites and among the three treatments. The first day of sampling, the results shows that the effect was due to soil type, only the second sampling which was after the month reflected the effect of treatments. The effect of treatments on soil moisture was proportional to the effect of vegetation, as well as the effect of soil type on soil moisture, thus vegetation production depends on soil moisture. Both stone lining and fencing off the land (excluding livestock) improved soil moisture levels and biomass production. The low-tech stone line is thus a possible appropriate technology for improving rangelands. Rotating grazing areas during the rainy season could also be beneficial.

Annex 1.18. An integrated evaluation of a small reservoir and its contribution to improved rural livelihoods

Basima Busane, L., Sawunyama, T.,L., Chinoda, C., Twikirize, D., Love, D., Senzanje, A., Hoko, Z., Manzungu, E., Mangeya, P., Matura, N., Mhizha, A. and Sithole, P. 2005. An integrated evaluation of a small reservoir and its contribution to improved rural livelihoods: Sibasa Dam, Limpopo Basin, Zimbabwe. In: Abstract volume, *6th WaterNet/WARFSA/GWP-SA Symposium*, Swaziland, November 2005, p32. [online]

The middle reaches of the Limpopo Basin in Zimbabwe contain some of the country's poorer communities, whose livelihoods are periodically threatened by recurrent droughts. One of the most common interventions by government and NGOs is construction of small dams - and a major construction drive is being initiated this year. An integrated case study of a small dam was undertaken at Sibasa in Insiza District to determine the characteristics of the reservoir that affects rural livelihoods.

Sibasa Dam, capacity 30,000 m^3, lies in the source of an ephemeral stream, which drains into the Insiza River. The average rainfall is 550 mm a^{-1} (40 year range 250 - 1100 mm a^{-1}). This dam is located above a fracture in the crystalline basement, providing perennial recharge to the dam from the aquifer. This has ensured that the dam never dried up since construction fifty years ago although variable rainfall has led to variations in its water storage, clearly visible from satellite images (1991-2005 range 10,000 to 35,000 m^3). Nitrate, phosphate, conductivity and hardness levels of the dam are in acceptable ranges for natural waters and WHO drinking water guidelines. Water in the dam is whitish, probably because of fine silt causing low light penetration. Phytoplankton at Sibasa is dominated by Dinophyta (90 %) with Chlorophyta, Bacillariophyta, Cyanophyta and Euglenophyta. The zooplankton is more diverse, but dominated by copepods (42 %). Plankton diversity is slightly lower than that of dams in the nearby Matopos National Park, which is less affected by human activities. The fish population is dominated by cichlids, providing a protein source for the community.

Although legally the dam is managed through the Upper Mzingwane Subcatchment Council, the local Chief controls access. He regulates the uses to which the dam may be put, restricting livestock during the wet season, when other water sources are available. The management system is not free of conflicts, however. The dam is supporting livestock, fishing, water for domestic purposes, and recharges a shallow borehole. Increasing the variety of water uses to support limited nutritional gardening could enhance livelihood contribution. Better use of existing dams, and improved catchment management may provide faster and more cost-effective returns to improving livelihoods than building new dams.

Annex 1.19.   The Green to Blue Water Continuum: an approach to improve agricultural systems' resilience to water scarcity

Vidal, A., Van Koppen, B., Love, D., Ncube, B. and Blake, D. 2009. The Green to Blue Water Continuum: an approach to improve agricultural systems' resilience to water scarcity. *Stockholm Water Symposium, World Water Week*, Stockholm, Sweden, August 2009. [online]

Farmers' reality around the world has always been to deal with a green to blue water continuum. Their dependency to this continuum has inspired them to innovate, and to extract the best productive value, not only from crops, but also from fish, livestock, and many other productive water uses. Through its research aiming to increase water productivity and to ensure more equitable use of water amongst users and the environment, the CGIAR Challenge Program on Water and Food has considered agricultural and natural resource systems as coupled to social ecological systems, and managed to increase the resilience of these systems through emphasizing not only the dynamics in each domain, but the nature and dynamics of their linkages.

This paper reviews two of CPWF projects in this perspective. One has been developed in a "green-water dominated" system, namely an endangered wetland of the Mekong basin, which has been shown to provide many direct and indirect benefits and services that are more resilient and less vulnerable to shocks than agricultural systems of various types and intensity occupying the same land-water interface, partly because it is fully adapted to and a product of the local ecological conditions resulting from the "flood pulse" phenomenon. The other one has been considering "blue-water dominated" Multiple Use Systems (MUS) in the Andean, Nile, Limpopo, Ganges and Mekong basins. It has shown that the water services ladder commonly used in the domestic sector failed to match reality in peri-urban and rural areas in low- and middle-income countries. Wherever water is available, people use water for productive purposes as well, including livestock watering, horticulture, irrigation, tree growing or small-scale enterprise. Ample and flexible choice among homestead-based activities accommodates volatile environments. Moreover, for women, the land-poor, and the sick, the homestead is often the only site where they can use water productively and in a resilient way. Both examples demonstrate that increasing water productivity and improving farmers' livelihoods should be done along the existing green-to-blue water continuum, and not only considering one side or another of this continuum, as still too often done.

About the Author

Born in 1975 in Lusaka, Zambia, David Love is a
Senior Consultant in the Groundwater and
Geochemistry Division of Golder Associates Africa.
He holds a BSc from the University of Zimbabwe
and BSc Honours and MSc from the University of
Stellenbosch. David is a Fellow of the Water
Institute of Southern Africa. During his PhD studies,
David worked for WaterNet, a SADC subsidiary
institution and a network building capacity in
Integrated Water Resources Management in southern
Africa, initially as Research Coordinator and later as Manager, and served as the
Chair of the Pan-African Steering Committee of the Partnership for Agricultural
Water in Africa (AgWA) and a Visiting Scientist with ICRISAT. David has lectured
at the University of Zimbabwe, University of Stellenbosch and UNESCO-IHE.
David is married to Faith, who is also a water and environmental scientist. They
have two children, Kathleen Taboka and James Robert Langanani. David and Faith
also farm cattle in the Mzingwane Catchment.

T - #0399 - 101024 - C262 - 240/170/14 - PB - 9781138001428 - Gloss Lamination